The decision makers: ethics for engineers

James Armstrong, Ross Dixon and
Simon Robinson

 Thomas Telford

Published by Thomas Telford Publishing, Thomas Telford Limited, 1 Heron Quay, London E14 4JD.
URL: http://www.t-telford.co.uk

Distributors for Thomas Telford books are
USA: ASCE Press, 1801 Alexander Bell Drive, Reston, VA 20191-4400
Japan: Maruzen Co. Ltd, Book Department, 3–10 Nihonbashi 2-chome, Chuo-ku, Tokyo 103
Australia: DA Books and Journals, 648 Whitehorse Road, Mitcham 3132, Victoria

First published 1999

A catalogue record for this book is available from the British Library

ISBN: 0 7277 2598 X

Typeset by The Midlands Book Typesetting Company.
Printed and bound in Great Britain by Bookcraft (Bath) Ltd.

Author biographies

J.H. Armstrong, *OBE, HonDEng, FREng, FICE,*
FIStructE, FRSA
Dr James Armstrong is a Past President of the institution of Structural
Engineers. He recently played a key role in redrafting the Rules of Professional Conduct for the Institution of Civil Engineers. He co-ordinates
the work of the Royal Academy of Engineering Visiting Professors and
he chairs their Design Matters Group. He is a Visiting Professor at
Kingston University. As a partner with the Building Design Partnership,
he was responsible for planning such major projects as the Channel
Tunnel Terminal works and the Falkland Islands Airport. He was awarded
the OBE in 1996 for services to both education and engineering.

J.R. Dixon, *BSc, CEng, MICE*
Ross Dixon is a Chartered Civil Engineer with 30 years' experience of
the construction industry. After training with a major civil engineering
contractor, he has worked with consulting engineering practices, holding
positions such as Resident Engineer, Chief Civil Engineer and Technical
Director of a Hong Kong office. He has broad experience of briefing,
feasibility studies, planning, design, construction and project management
of building and civil engineering projects in the UK, Bahrain, Saudi
Arabia, Qatar, UAE, Philippines and Hong Kong. He is currently a
Senior Lecturer in the Construction Management Group of the School
of Civil Engineering, University of Leeds.

Reverend Dr S. Robinson

The Reverend Dr Simon Robinson is Senior Anglican Chaplain to the University of Leeds and a Lecturer in Christian Ethics. He is a member of the Centre for Business and Professional Ethics at the University. Initially, a psychiatric social worker, he was ordained in 1978 and, after curacies in the Durham diocese, he moved to Heriot Watt University as Anglican Chaplain, teaching moral philosophy and working closely with the Department of Civil Engineering in the development of professional ethics courses. His interest in civil engineering has continued at Leeds, where he and Ross Dixon are preparing a ground breaking course on engineering, management and ethics. His publications include *Serving society: the social responsibility of business* (Grove), and several papers on business and professional ethics. He is in the process of writing a book on case work and business ethics, and another book on spirituality and health care.

Preface

There is concern today about standards of conduct in life generally. One feature of that concern is the proposition that British society may be moving not towards immorality, but towards amorality. There could be no more appropriate time for the engineering profession to re-examine and to re-state its own code of ethics. This is what the Institution of Civil Engineers has been doing, resulting in a revised code of behaviour for its members.

This book by James Armstrong, Ross Dixon and Simon Robinson draws on this recent work by the Institution and describes the changes — but it is far more than a report on these developments. The authors tackle the subject of ethics from its roots in the major philosophies and religions of the world, through the trunk and branches of regional and professional differences, to the fine twigs and leaves of how the engineer should approach his or her work and respond to ethical dilemmas.

Ethical dilemma is perhaps a keynote of the book, for there is a continuing and justified emphasis of a truth which all experienced engineers will recognise. Ethics in the modern age is rarely, if it ever was, a choice between good and evil; ethical issues are painted in shades of grey; they lack the definitive solutions that are available in nearly all cases of engineering analysis. For that reason, teaching and training in this area present a particular challenge which the book most usefully discusses.

Some wag once defined ethics as 'a county south of Suffolk'. It is not. It is a very cornerstone of our profession. I commend this book to any engineer who wishes to have a better appreciation of that cornerstone.

Roger Sainsbury

Abbreviations

ASCE	American Society of Civil Engineers
BS	British Standard
CDM	Construction (Design and Management) regulations 1994
CPD	Continuing Professional Development
DFBO	Design/Build/Finance/Operate (contracts)
EC	Engineering Council
EDC	Economic Development Council
EEC	European Economic Community
FE	Further Education
FIDIC	Fédération Internationale des Ingénieurs Counseils (International Federation of Consulting Engineers)
HE	Higher Education
HLV	High Landscape Value (area of)
HNCV	High Nature Conservancy Value (area of)
ICE	Institution of Civil Engineers
ISO	International Organisation for Standardisation
IT	Information Technology
LHDA	Lesotho Highlands Development Agency
MNC	Multinational Corporations
PFI	Private Finance Initiative
PSA	Property Services Agency
QA	Quality Assurance
RSA	Republic of South Africa
SSI	Special Scientific Interest (area of)
TQM	Total Quality Management
UN	United Nations
WFEO	World Federation of Engineering Organisations

Contents

1

The ethical field

It is the duty of the civil engineer to harness the great forces in nature for the use and convenience of man

ICE Charter

1.1. Introduction

We all make decisions — but civil engineers make big decisions. We can create greater well-being, or longer-term problems, than most prime ministers or presidents. The Roman road builders have had a longer lasting influence upon the shape of society than their Emperors' wars. Irrigation projects built 1800 years ago in northern Sri Lanka are still serving the inhabitants faithfully. The traditional Nile Delta fisheries will never recover from the change in regime brought about by the Aswan Dam project. Many of us marvel more at the construction challenge of the pyramids than ponder on their spiritual significance. The ubiquitous engineer-designed car and computer are bringing about more social changes than the National Health Service. Engineering decisions affect the quality of life of large numbers of people throughout the world.

When making a decision, an evaluation of alternative courses of action is required. Evaluation requires a set of basic values to use in the comparison of the relative merits of the nature and outcome of such alternative actions. This book explores the problems of decision making by the professional engineer. The intention is to present the individual, family, community, and global contexts within which professional engineers have to make decisions, and to illustrate these by reference to specific situations — large and small, general and personal — in which professional engineers find themselves. An attempt is made to establish a framework against which we can conduct an 'ethical audit' of our activities, showing the changing ethical factors that need to be considered in differing situations and at different stages in the development of a project.

> *A promising young engineer on your staff refuses to accept a valuable overseas posting — he is just about to get married. What do you do with him?*
> Send him somewhere else if he really is that promising!

Ethical decisions are not just decisions about the best way to meet a given brief or objective. They are concerned specifically with the quality of our decisions — with justice, with equity, with the consequences for all affected by the decision, and with the personal and collective responsibilities which lie beyond the contractual obligations into which we enter. They are concerned with the 'good' and the 'right', with conflicts between rival goods or ills. We each develop a personal value system influenced by the social and cultural norms of our environment. These systems change continually as exposure to ever-wider experience modifies our attitudes.

We live in an age when technical and political changes take place more rapidly than at any time in the past. The scale and range of societies, the size of projects undertaken by them — projects concerned with education, health care, government, public and private works, and many other areas of communal concern — have greatly increased over those undertaken in previous centuries. Perhaps the rate of change is even more significant than the scale of change.

> *A responsible public sector engineer is a client's representative on a major project. He is also the chief executive of a wholly-owned subsidiary company providing testing facilities. Without his knowledge his company bids and secures the testing contract on his own project. What should he do?*
> Resign from the Board to ensure no conflicts of interest.

In parallel with this development there has been an unprecedented increase in the population of the world. Changes in the demographic distribution of the population with respect to the age spectrum, require quite different social relationships and dependencies than has previously been the case. These factors affect our traditional values and require a reappraisal of our concerns with the nature of morality, and of our individual and collective responsibilities.

In addition to the scale and rate of change, and the changes in population distribution, we have to recognise the nature of those changes. Technological developments have placed vastly increased sources of power under the direction of mankind, and have greatly enhanced the speed with which we can communicate with one another. Within decades the time taken to communicate with some parts of the world has been reduced from months to seconds. These developments have resulted in the need to recognise that world communities are becoming much more interdependent. Differing views as to the nature of the 'good' and of the 'right' have a very real impact upon the peace and prosperity of all the peoples of the world.

Increasingly, the decisions of a single powerful individual can affect the lives of many millions of others — for better or for worse. The role of the modest and responsible engineer is more powerful than we think.

The contemporary ethical scene is one of confusion. In the so-called 'post-modern age' there is no single vision or definition of the 'common good'. Instead there is a stress on liberalism (the individual can think or do what he likes provided that it does not interfere with the freedom of another) and pluralism (the acceptance of many different views about what is good). Social commentators note that this gives the appearance of moral breakdown, especially when the 'grand narratives' on which were based so much of the sense of duty or moral obligation — such as the role of the British Empire before the First World War or the role of the family — have either been lost or radically changed.

Such a complexity of views can cause us to move from the anarchy of extreme liberalism to the totalitarianism of complete centralised control (see Fig. 1.1). Ethical judgement is needed to maintain the stability of society without the undue imposition of formal regulations for every act. With sound judgement it is possible to maintain the 'sovereign's' peace without such excessive regulations — discrimination reduces the prevalence of crime.

Both the general public and professionals declare an interest in this area, but statements based, no doubt, on good intentions, are frequently made which display unsound reasoning. Engineers are concerned with balanced judgements, requiring understanding and independent evaluations. They need to know which decisions are likely to produce the most favourable results, both in the short-term and in the longer-term.

The science of 'ethics' addresses issues having both moral and practical significance. Decisions are not simply reactive, such as choosing which

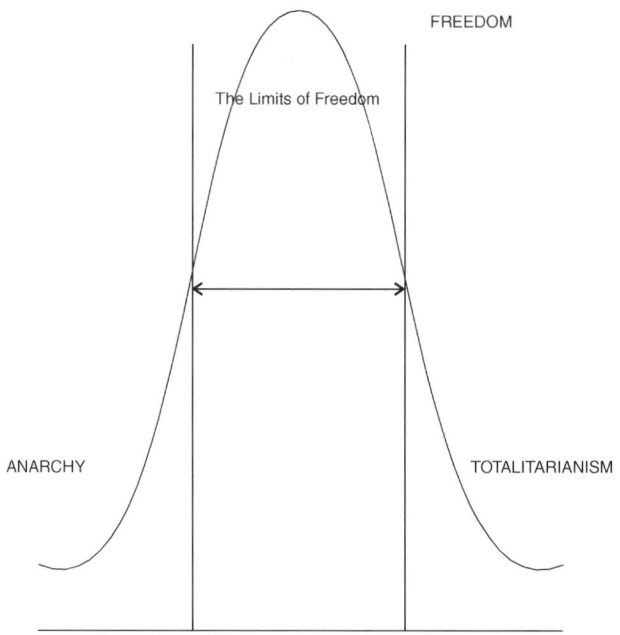

Figure 1.1. The Limits of Freedom

bus to catch, but involve concepts of duty and responsibility, and of conflicts between rival goods or ills. We each develop a personal value system influenced by the social and cultural norms of our environment. These systems change continually as our exposure to ever-wider experience modifies our attitudes.

Reflections upon the ethical significance and correctness of any situation require the use of judgemental reason which is not always familiar to engineers, fine-tuned as we are to the use of deductive logic and the pursuit of the 'correct' answer to any technical problem.

> *A much loved private house, on the access route to a major beneficial development, must be compulsorily purchased and demolished if the project is to proceed without considerable delays and extra costs. The alternative route would remove the habitat of a rare slipper orchid, and cost an extra £1 000 000. How important is the owner's personal happiness relative to the loss of the flora?*
>
> It is almost certainly possible to compensate the householder, but not possible to regenerate the primitive habitat.

Ethical problems rarely have a single readily deduced solution. They deal in shades of meaning, uncertainty, doubt and complex relationships. They are part of an ethical continuum whose components are inextricably entwined with one another and where a view is needed of the whole. There is an indivisibility in the continuum. Personal values cannot be separated from societal or professional values, although it may be possible to reflect upon particular aspects to develop a comprehensive moral attitude. It is not possible to deal in precise rules and regulations, only to suggest guidelines. These guidelines give ethical behaviour its special place in maintaining the quality of life in communities, and distinguishes it from lawful behaviour, which follows positive instructions evolved by a society to curb defined antisocial behaviour. The distinguished philosopher Lord Wolfenden once said, '*The essence of ethical behaviour is not the conflict between Good and Evil, but the conflict between rival goods*'.

However, the enjoyment of reflective moral thought is very real and gives a freedom of choice which is refreshing after the formal disciplines of rigorous analysis — which gives a different kind of satisfaction. The human being is blessed with the ability to enjoy and develop both modes of thought, as well as the ability to appreciate beauty through the aesthetic sense.

In considering the areas of decision making particular to civil engineers, with the international dimensions and the scale of influence of those decisions, we shall relate them to our personal systems and to the context of national and international laws and conventions. Case studies are used to demonstrate the underlying principles of responsible and humane ethical actions.

The book is structured to give a systematic presentation of the nature and context of ethical actions and their relationship to particular types of engineering problems — covering such issues as human relationships, competence, responsibility, cross-cultural issues, environmental issues and financial issues.

1.2. The general principles

The general principles, with some of the underlying philosophy, are set out in Chapter 2. The nature of the 'good' and the 'right' are discussed. The relationships between ethical practice and national and international laws and regulations are considered alongside problems of cultural diversity, conflicts of interest and specific major ethical questions. Problems are cited to illustrate the range of these ethical principles.

The development of personal and societal value systems is also discussed in Chapter 2. It is brought together with some discussion on the range of consequences of our decisions, to form a map of the decision maker's world. Chapter 3 focuses on the specific place and nature of professional practitioners and their responsibilities towards, and service to, humanity.

1.3. The ethical audit

As a structure for exploring the wide field of civil engineering ethics, the stages involved in the development and procurement of a major project are set down in Chapter 6. The stages vary somewhat in detail from project to project, but the general framework is remarkably similar in principle, even if the contractual and physical problems are very different.

Figure 1.2 illustrates such a structure. The major stages in this hypothetical framework are covered in detail in Chapter 6. In abbreviated form they are set out below.

- People — the natures, responsibilities and aims of the several interested parties — including special interest groups as well as the principal decision makers.
- Project brief — the content and definition of the brief for a project.
- Context — the social, political, economic and legal parameters of the project. The topographical considerations, resources available, services needed, and the provision of the necessary labour, materials and equipment.
- Proposals — the development of alternative solutions to the brief within this context, and the appraisal of these solutions against agreed standards of performance, cost and other factors.
- Implementation — the organisation and management of the design, development, construction and control of the implementation of the project, including labour and material policies and public relations.

In illustration of this approach we have considered a major project. Following the conflict in the Falkland Islands it was decided to improve the airport facilities on the Islands to enable long-haul jets to fly non-stop from the UK. There were many factors to be considered at every stage of the project. They all impacted upon the quality of life of all the parties concerned, and upon the quality of the environment in the Islands. A fully worked study of this project is included in Appendix 1.

The use of the suggested framework shows how the situation can be assessed and the principal factors identified at each stage, taking into account the different value systems of the parties concerned and the conflicts of needs that inevitably arise when we set out to modify our environment in some way.

1. The decision makers

Notes on parties involved — decision makers, advisers, other parties

Interdisciplinary team working

Authority — extrinsic/intrinsic

Motives, aims, objects

Conflicts — divergent, convergent negotiations

2. The brief

Wants

Needs

Constraints

3. The context

Societal context

Physical context

Constraints

4. The project design

Long term effects — designer responsibilities

Reversibility — designer/owner responsibilities

Sustainability — all responsible

Maintenance — owner responsibilities

Environmental impact assessment

5. Implementation

Team structure and procedures

Control systems, monitoring, quality assurance, CDM regulations

Construction issues

Programme

Labour and material policies

Equipment

Figure 1.2. The project ethical audit

1.4. Other factors

In addition to setting out the framework for an ethical audit and showing its application to a specific project, chapters are included that suggest ways of teaching ethics to engineers. Other chapters discuss major ethical problems such as corruption, contributions to public policy issues and cross-cultural conflicts of customs together with further illustrative case studies and procedures for conducting ethical audits.

2

General principles

2.1. Introduction

Sometimes engineers take pride in the knowledge that they are decision makers, and sometimes it terrifies them. Either way, they cannot escape the responsibility. 'I did what I was told', is no excuse for immorality any more than it is a legal defence. In exploring the problem of decision making, particularly from the viewpoint of the civil engineer, some of the fundamental principles of ethics — alternate viewpoints, moral attitudes, the 'golden rule' acknowledged by many differing cultures, the concept of 'moral distance' — will be examined.

We then explore the decision maker's world (see Section 2.3). This world can be considered as three neighbouring and interdependent frameworks — our personal value framework, the framework of the societies in which we operate, and the range of consequences of our decisions.

> *To be is to have duties.*
> Sydney Evans
> Former Dean of Salisbury — 1980

The privilege of self-consciousness, carries with it the obligation to fulfil certain duties. Those duties are variously formulated by different cultures and civilisations, even by different social groups in the one society. The Charter of the Institution of Civil Engineers (ICE) is one secular formulation of such fundamental principles, if the 'use and convenience of Man' is extended to include the service of Creation as a whole.

Ethical meaning can be considered through the concepts of duties and rights. It has been said that duties are axiomatic, while obligations are working hypotheses.

- Duties are offered unconditionally. They tend to be promissory and open-ended, such as the general duty owed to humanity.
- Obligations are based upon contractual relationships and are specific. Non-fulfilment of the obligation can lead to the end of the relationship.
- Rights are basically freedoms which enable the individual to fulfil important human needs such as family support or education. The precise list of human rights is a matter of debate.

Traditionally, rights are divided into moral rights, argued for on the basis of need, and legal rights, which are recognised and protected by law.

This raises questions about the relationship between morality and law. The two are closely related, as would be expected, since they share key concepts such as justice, rights, rules and responsibility. The two areas are logically distinct. Law and legal justice are defined in terms of the procedural rules of the legal system. The law does not decide what ought to be legal or what the criteria of responsibility ought to be. Morality can provide a critique of the law and can argue for ways in which the law might be changed or interpreted, but the law cannot determine morality. Much greater flexibility is possible in an ethically sound society, in which just exchanges can take place without the need for restricting regulations, providing always that the mores are such that an excess of liberalism does not turn to unbridled licence.

Ethical decisions are concerned with the 'goodness' of things, or the 'rightness' of actions. Engineers' decisions have both a moral and a practical significance. There are decisions which are almost entirely practical, like opening an umbrella in the rain, or filling up the car with petrol. The concepts and decisions discussed in this book are not simply reactive, but are concerned with actions involving concepts of duty, responsibility, the 'good' and the 'right'. Should I adversely affect this natural habitat? Should I use these fossil fuel resources on this transient project?

> *Duty is the action proper to each man which keeps to what is fitting and honourable as circumstance, person, place and time require.*
> Marsilio Ficino, Renaissance philosopher — c.1450

The beautiful complexity and flexibility of the interaction between humanity's aspirations and the generous provisions of nature give rise to many areas of apparent conflict, requiring decisions to be made related to the quality of life for many thousands of people, perhaps for generations. Such decisions require an ethical understanding equal in rationality and competence to the technical understanding needed to cope with the physical dimensions of an engineering project.

Typical of these complexities are

- 'rival goods' — conflicts of interest
- cultural diversity
- environmental issues
- sustainability
- national interests, political/economic/security
- trans-national projects
- conflicts of law — justice
- contracts — competition
- private v. state finance
- design build
- financial investment policy — initial cost v. operating costs
- employment
- health and safety.

The formulations of what things are 'good' and which actions are 'right', vary considerably from culture to culture, and from age to age. The age-old and continuing attempts by philosophers, theologians, idealists and others to arrive at some fundamental formulations have proved difficult, and the results of these attempts are sometimes very obscure. In the meantime we must do our best by society and the environment to raise the quality of life and not to damage seriously and permanently the ability of the world to continue to support future generations. As engineers we have to 'get on with it'.

2.2. What is ethics?

2.2.1. Alternative viewpoints
Views differ as to what is good. The study of ethics can help in two ways.

Firstly, ethics involves the discipline of systematic enquiry into moral norms or standards of behaviour and their underlying values and justification. It can help us to understand and justify these values.

Secondly, applied ethics looks into the ways in which moral value (that which is valuable for the good of the environment and its inhabitants) can be applied to particular areas of concern such as the professions or business. This includes handling the very different ethical claims that may come out of different groups in any situation. Applied ethics reminds us that while the discipline of ethics may be treated as separate from other disciplines (such as technology or economics) when it comes to practice, the ethical dimension is a fundamental part of all judgements and decisions. What becomes important then is the development of 'ethical thinking' which is a part of the decision making process.

Definitions of the moral good can be based upon one of several philosophical theories.

- Deontological (to do with duty). Deontological theories see good as something which is self-evident and which can be summed up in general rules which apply to all: e.g. it is wrong to kill. In this theory the action of killing is always and intrinsically wrong. It breaks the basic rule of respect for life. Moral good is discovered through reasoned reflection. This approach is concerned primarily with the quality of an action, with the means of achieving a desired end.
- Consequentialist (concerned with outcomes). Such theories see good as defined not by the action, but by the consequence. In this theory, killing another human being may be morally good providing that it gives rise to some greater good. For example, it has been argued that the deaths caused by the first atom bomb were morally acceptable because the consequence was to save far more lives through the consequent earlier ending of the war. One particular form of this theory, utilitarianism, sees the good as being defined by the maximisation of good consequences — the greatest good for the greatest number. This theory argues that we can know what is good only when we have fully understood the context.
- Virtue (concerned with moral excellence). From this viewpoint good is found neither in the act nor its consequence, but in the character of the moral agent. Such an approach stresses that no rule can give guidance in every situation and that we cannot effectively weigh up all the consequences. Thus we need to develop the capacities to deal with each situation as it arises. This demands the development of personal or professional virtues, qualities which enable the person to *be* good and from this to *do* good.

All of these theories have their flaws and in the end cannot exclusively define what is morally good. All things being equal, certain actions are self-evidently wrong. However, this cannot be applied rigorously in every situation. Moreover, the deontological theories assume the reasonableness of human beings, i.e. that they are capable of seeing what is good. Ironically, deontological theories rely upon consequentialist considerations to justify themselves. Respect for life, for instance, is justified ultimately by the survival of the self, the community and even the species.

The consequentialist approach ignores the importance of intention and motive in any ethical judgement. It also runs the danger of the ends justifying any means and even of denying the rights of minorities. In deciding between different options there has to be some understanding of what constitutes the good in the first place.

The principles implicit in the concept of individual virtue rightly emphasise the importance of not simply deciding what is good, but of actually achieving good ends. Any virtue itself depends upon some prior definition of what is good.

All three viewpoints are necessary for an understanding of what is good in any situation, to obtain a 'fix' upon the truth in a real-life situation. The practical ethical enterprise has to develop a methodology which enables all the above factors to be taken into account.

2.2.2. The fundamental moral attitude

Ethics is not about simply recognising an objective good, not least because ethical knowledge is not objective in the way that empirical or scientific data can be. We rather bring our ethical meaning and judgement to a situation and try to organise relationships and groups on the basis of that ethical perspective. At the base of this is a concern of one for the other which might be termed the fundamental or basic moral attitude. Immanuel Kant sums this up in the idea of respect for persons. He argues that it is our duty to respect persons and that this involves, 'acting in such a way that you treat humanity . . . never simply as a means, but always at the same time as an end'.[1]

At the heart of this idea of respect is the recognition of common humanity. In this, the other person is seen as being essentially the same as the moral agent, of equal value and having the same needs. This may be summed up in terms of the 'golden rule' — a concept common to many different ethical systems and cultures, including the following.

- *Christian version.* 'Treat others as you would like them to treat you.' (Luke 6, v.31) 'Love your neighbour as yourself.' (Matthew 22. v.39)

---

[1] Kant I. *Groundwork of the metaphysics of morals.* (Translated by Paton H.J.) Harper and Rowe, London, 1964, p.70.

- *Hindu version.* 'Let not any man do unto another any act that he wisheth not done to himself by others, knowing it to be painful to himself.' (Mahabharata, Shanti Parva)
- *Confucian version.* 'Do not do to others what you would not want them to do to you.' (Analects, Book xii # 2)
- *Buddhist version.* 'Hurt not others with that which pains yourself.' (Udanavarga, v.18)
- *Jewish version.* 'What is hateful to yourself do not do to your fellow man. This is the whole Torah.' (Babylonian Talmud, Shabbath 31a)
- *Muslim version.* 'No man is a true believer unless he desires for his brother that which he desires for himself.' (Hadith Muslim, imam 71–2)

In these versions of the 'golden rule', there is a balance of concern for the self and the other, which is in essence unconditional. It does not depend upon any quality or group or type of person. In general the balance of the 'golden rule' reminds us that the needs and values of the individual have to be taken into account as well as those of others.

There are several implications to this principle of individual needs and value.

- It is impossible to build an adequate universal ethic on the interests of any particular section or group.
- The nature of morality has to transcend such interests not least because if it were based upon such interests it would lead to exclusion and injustice.
- Morality must always start with an acceptance of responsibility both for the self and for the other.

Such a basis of morality is not simply rational. The essence of morality cannot simply be captured in any calculus or any rules or codes. Rules and calculation cannot be avoided when the details of an ethical decision are worked out. They give good guidance and are helpful in trying to weigh up different options, but they cannot provide the essence of the ethical response in any situation.

Strict adherence to rules can, moreover, actually lead to a denial of responsibility. A good example of this is the *Herald of Free Enterprise* ferry disaster. The Sheen Report[1] notes a failure of responsibility at several levels, despite a clear code of practice. In this case the individuals

---

[1] Department of Transport. *The Sheen Report.* Report of the court, No. 8074. DoT, London, 1987.

in question chose not to go beyond *their* role in the code, even though there was opportunity for them to close the bow doors, and thus they did not recognise a wider responsibility.

Behind the denial of responsibility is what is sometimes referred to as 'moral distance'. 'Who is my neighbour?' This is of particular importance to civil engineers whose decisions may affect the quality of life of generations to come, located thousands of miles away from the prime seat of actions. Many decisions have environmental and ecological consequences which may not immediately affect any of the primary actors in a project, but which still require wise ethical judgements.

Responsibility towards an 'other' who is mute and defenceless — such as the environment — requires particularly high standards of professional integrity and objectivity. Civil engineers may sometimes find themselves acting as the 'friend in court' on such issues.

This 'moral' distance may lead a person either to be unaware of the others involved in the situation or to deny the humanity, and thus the moral claims, of the other. Such moral distance can be encouraged by the way in which social groups are organised. For example, the effect of the division of labour, which emphasises responsibility for only a limited task; the rise in bureaucracy, with the attendant pressure to see human beings as 'cases'; the rise of the expert, with the concomitant concern for the practice of specialised skills and of technical solutions to problems. The tendency for these things to lead to 'dehumanisation' is exacerbated by other pressures such as fear of job loss, and can lead to conflict between sharply defined rival sectoral 'goods'.

Four principles have been suggested as a reference point for the professional; these are

- equal respect for the autonomy of others
- concern for the good of others
- a concern to avoid harming others
- justice — a concern for equitable treatment for all parties.
  (Beauchamp and Bowie[1])

The basic moral attitude, referring to these principles, is concerned with our approach to all situations. There must be a concern for the safety of the other, but such a concern cannot encapsulate all the moral requirements, not least because one way of keeping another safe is to deny freedom. Equally, there needs to be a concern for justice in any

---

[1] Beauchamp T.L. and Childress J.F. *Principles of biomedical ethics.* Oxford University Press, 1989.

relationship, but the concern for the particular good of the other ensures that people cannot be treated in a uniform way.

This provides a basis which is both principle and attitude. The principles set an irreducible standard with certain values which have to guide any ethical interaction. At the same time it provides a moral attitude. Such an attitude radically affects how we view reality.

2.3. The decision maker's world

> *We should live honestly, should hurt nobody and should render to every one his due.*
> Justinian

As living and involved members of the world our decisions are a complex mixture of our individual natures, of the wishes and customs of the societies in which we operate, and of our awareness of the consequences of those decisions not only to ourselves, but to others and to the environment as a whole. Therefore in mapping this world, we cannot ignore any of these factors. They are like the three dimensions of space — each penetrates and influences the others.

A model is presented which shows the relationships between these three aspects of our world. Each of them has a profound and all pervasive influence on the quality of decisions made at every stage in the evolution of an engineering project. They are set out in diagrammatic form in Fig. 2.1.

As moral decision makers we are concerned primarily with man's activities in the world of the mind — it is usually too late once the action is running! It is in the mind that we formulate our understanding

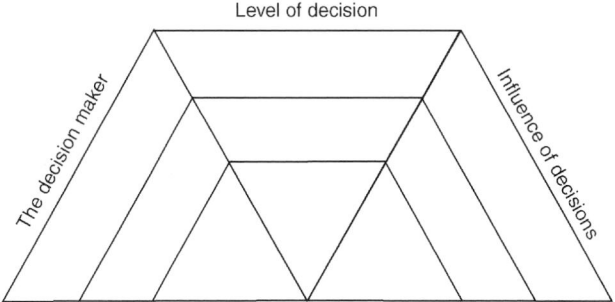

Figure 2.1

of the societies in which we live, and in which decisions affecting all our voluntary actions are made. Many decisions made at this level are purely logical or analytical in their nature, uninfluenced by considerations of good or bad. As ethical members of a society we are concerned with those decisions and beliefs that influence the moral quality of our lives, which enhance or detract from our ability to realise our full potential and which contribute to the well-being of others.

2.3.1. Personal value systems (Fig. 2.2)

Our personal 'conditioning' arises out of our individual upbringing and culture — our family background and our basic education. We learn from our parents and relatives, from friends and teachers, certain patterns of behaviour and certain concepts of what is considered to be good or bad. We also learn which actions receive the approval or disapproval of our social group. The process of learning is frequently one of trial and error. We have all observed how small children explore their relationships with others, testing the limits of what is and is not permissible.

By the time we have reached adulthood we have acquired some relatively well-formulated opinions concerning our behaviour and the behaviour of others. These opinions reflect the attitudes of the society and of the family in which we have lived.

In addition to the opinions acquired from his family and company, each human being has individual talents and aspirations. These individual characteristics, whilst expressing themselves through the customs and habits of a particular society, are related to more fundamental human attributes.

Consider, for example, the talents for music or mathematics. These talents have been evident in members of all races and at all times in recorded history. They clearly transcend national or cultural values, and are not the properties of a particular age.

Unless this is appreciated, the talent itself can become lost in a particular local expression of that talent. When the universality of talent is more fully appreciated it can be found to be very helpful in developing relationships with others of different nations and cultures, and in understanding the writings and actions of men and women from different historical periods.

In parallel with these individual but universally evident talents, we can consider individual, but also universally evident, motives. Again, the power of the motives behind our actions is normally expressed through the common usages of our social context. However, human motives are more fundamental than human cultures, for there will always be men and women who seek power or wealth, or who wish to serve or teach. Such motives are universal.

Where this is appreciated the individual finds that his wishes can be pursued in any context. Clearly our individual natures are influenced by these personal qualities; qualities of talent and motive, and these are placed in the second level in Fig. 2.2.

Every individual sees the world from a different viewpoint and these differences frequently lead to misunderstandings. Our personal and social mores are likely to differ significantly from those of others acquired within different societies, perhaps with very different cultural and ethnic traditions. Major differences are usually very obvious and, paradoxically, more acceptable, than are the minor, but significant, variations that can exist between regional or family groups. By appreciating some of our own attitudes we will arrive at a better understanding of our position in human society and in the world in general.

The factors that we have discussed thus far as contributing to the nature of the decision maker do not necessarily produce what we might call ethical man. He might be courteous, honest, clean and articulate. Raised in what might be considered as a 'good' context, the individual may acquire good habits. He or she may have combined a talent for speech and with an individual will to lead others, becoming an effective manager or director of affairs in society. But such a person may be no more virtuous fundamentally than the individual who has acquired bad habits, and whose manipulative skills, combined with a love of wealth, have turned him into a master thief!

That which distinguishes the ethical person from the well-behaved is a belief in values. The moral maturity that comes with full development brings the capacity to take responsibility for ethical decision making.

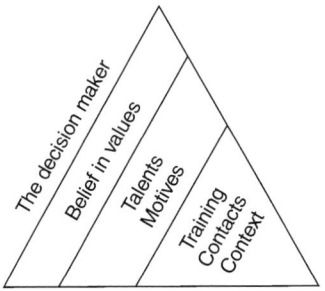

Figure 2.2

Responsibility makes freedom possible, preventing anarchy and undue licence.

For the decision maker to be a moral being he must believe that values are part of the order of creation, even if they have not yet been clearly formulated or fully understood. It is useful to revisit our personal beliefs from time to time, to consider whether these are fundamental or merely relative and habitual.

2.3.2. The virtues

The development of a personal value system results in 'moral maturity' — a state in which the individual has developed the capacity to take responsibility for ethical decision making. It involves a moral freedom, the freedom both to see and to do one's duty. This implies a 'virtuous' person.

The importance of the virtues in ethical practice is clear. It is said that there are four fundamental virtues (Plato's 'Divine Virtues'). They are

- Wisdom
- Temperance
- Fortitude
- Justice.

Wisdom
Wisdom is an openness to the truth in all aspects of existence. It is closely related to the skills of listening and being still. It is crucial for avoiding premature judgement, and for achieving creative planning.

Temperance
Temperance is essentially the quality of moderation, balance and self control. It does not preclude having fun, but simply ensures that the person is not controlled by passions such as anger or fear, which tend to get in the way of good judgement. It is good to be well or even-tempered.

Fortitude
Fortitude looks to the qualities of courage and resilience. It enables a person to make better decisions and to stand against the pressures and challenges which might corrupt judgement.

Justice
Justice is the quality of even judgement which enables equal treatment of all parties. It requires an awareness of right and wrong, and hence the acceptance of some code of standards.

The different virtues are interrelated. There is an essential unity of the virtues. We may perhaps have one virtue more fully developed and understood than the others but there is no doubt that all are essential to each other. Honesty is important for faith or trust to develop, fortitude for temperance, and so on. The concept of teamwork can be built upon the virtues of the different members, accepting that it is rare to find someone with all the virtues fully developed.

Whilst individual value systems are clearly different, the development of moral meaning is essentially a social process involving continual reflection and development based on experience and the accepted norms of a particular society. These range from the ethos of school and work to the ethos of intermediate organisations such as the church, professional organisations and social groups. These are all related to more general and cultural value systems.

2.3.3. Societal value systems (Fig. 2.3)

If we want the freedom to do as we like, when we like, then our behaviour will be anarchic and our ability to benefit from the skills and talents of our companions will be much reduced, or denied completely. The anarchist has no real freedom.

If we wish to be directed from the upper levels of national government in all the details of our transactions, then we shall lose our freedom to innovate — this is the totalitarian approach, the command economy — and again there is no freedom.

The degree of freedom necessary for the full development of man, as stated in Chapter 1, lies somewhere between anarchy and totalitarianism and is a measure of the maturity of the society, and of our 'freedom under

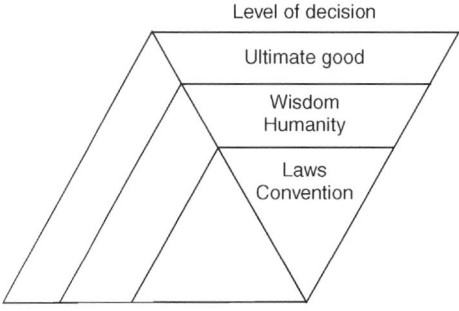

Figure 2.3

the law'. Plato described the object of laws as 'to make those who use them happy and to confer every sort of good'.

In the nature of things, hierarchies arise in societies, reflecting, in healthy communities, the different levels of understanding of the members of those societies, and their consequent different responsibilities. A description of such a hierarchy is given in Table 2.1. It sets out the

Table 2.1 A natural hierarchy

Level 7	Humanity	Teachers, prophets, master philosophers
Level 6	Civilisation — lawgiver	Men working at this level are few and far between, for they give laws upon which civilisations stand — such as the fundamental precepts of the great scriptures of the world.
Level 5	Culture — understanding	It is important that we see this as a function rather than as any individual person. People of understanding are those with faith and vision and the quality of their work is such as to strike a chord in the hearts and minds of others, to fire an allegiance which can last for centuries.

The economic structure

Level 4	Capital city — government	The level of government acts as a bridge between the economic organism and those from whom the basic ideas and laws in society come. The term government includes the church, universities, judiciary, institutions of the sciences and the arts, etc.
Level 3	Regional commercial centres and ports	Wholesale markets, financial institutions, large-scale manufacture, main ports, centres of commerce as distinct from general trade — all those activities which link the centres of trade nationally and internationally.
Level 2	Market town	The level at which goods and services are exchanged. Within reach of the general population. The centre for local crafts and the retail trade.
Level 1	Production	The first level of communal life. The husbandry of land, crops and livestock, fishing, mining, building, etc., all activities concerned with our basic human needs, working directly on the natural resources of the creation.

functions discharged by people at all levels in society. The existence of such a structure, given communication and understanding between the levels, provides the freedom and the means for further development by all members of society. Some of our disputes arise because of confusion between the respective roles being played at each level.

> *You have dismissed an employee for idleness. Someone writes asking for a reference for him. How do you reply?*
> You must describe his attitude to work, perhaps with suggestions as to how it might be improved.

Considerations of the nature of society show how, through working together, men can increase their freedom. To gather the talents with which we were born and apply them to the knowledge available to us, in the best possible way, to meet a real need — that is real work. This is 'wealth creation' in the fullest sense of the phrase. This is real 'added value', the measure of the work that has been done. Human nature enables us to work within a hierarchical social system, extending the benefits arising from specialisation and from greater understanding, to all the members of a society. At each level in the hierarchy we can see how the responsibilities of those engaged in working at that level increase.

Collaborative societies allow us to exercise our skills and talents in some specialised way, to serve others with those talents. In this way the 'added value' is further increased. The resource created by that work can be used as society chooses. A group of people working willingly together can achieve much more than the sum of their individual efforts. Service to others is an essential part of social life.

Each level in a society depends for its ordered activities upon the stability of the level above. Thus the tradesman requires a stable currency, the merchant requires peace between nations, governments should draw their principles from men of understanding, and so on. These established values within a society form the moral context within which the members of that society are free to exercise their individual talents and desires. Corruption or conflict at the higher levels inevitably spills over into transactions at the lower levels. The influence of these social structures upon our value judgements is significant.

> Law
> *All are equal before the law and are entitled without any discrimination to equal protection of the law. All are entitled to equal protection against any discrimination in violation of this Declaration and against any incitement to such discrimination.*
> Universal Declaration of Human Rights - Article 7

Our decisions are influenced by the customs and laws of the societies in which we live. These are usually generated over many centuries, if not millennia. Some of them are enshrined in the culture, others are defined in the laws or statutes of particular nation states. These need to be considered carefully when making any major decision likely to affect a significant number of people. These social patterns are an essential part of our behaviour. Without them beneficial social exchanges would not be possible.

The codes of professional behaviour established by the major professional institutions are an important element in the 'informal' governance of nations. This is particularly significant with the greater interactive international dimension to relationships evident today, which is much more difficult to regulate by law.

In any society, divisions of labour arise as the members of the society develop their mutually supportive separate activities. Because we seek to work with one another — indeed must work with one another if the quality of our lives is to be maintained and enhanced — we agree to perform certain tasks, receiving something in return for our contribution. In a complex society we rarely receive goods or services directly from those to whom we make our goods or services available. Barter permits only a very limited amount of interchange, and only over short periods of time. Thus arises the means of exchange and of credit, which we now take so much for granted, but which depend so much for their effective operation upon the moral climate and standards of the members of the community.

> Convention
> *Methods of measurement of total or effective thickness of thin surface hardened layers in steel (produced by treatments such as shot blasting, shot penning, flame or induction hardening, carbonitrading, and carbonising and hardening).*
> British Standard 6286

Some of these exchanges are not only remote in organisation — like the purchase of some foreign manufactured goods — a Japanese car, for example, with long supply chains of component manufacture, distribution and assembly — but also in time. For example, the delivery of an offshore oil rig in three years time, or the underpinning of someone's garage when they are away from home in six months time.

Depending upon the situation of the moment, we operate at different levels shown in the hierarchy diagram (Table 2.1) at different times. In our daily lives we all have, from time to time, to perform simple manual tasks, whether these be washing up cups, emptying ashtrays or digging in the garden. These are all performed at level one of the hierarchy diagram. More complicated tasks, requiring interaction with other members of a society — commerce, management, education, etc. — are also easily recognisable.

In Table 2.1, levels 1 to 4 relate to the village, the market town, regional centres and government. These four levels are all contained usually within one separate sovereign state which establishes certain standards of behaviour and styles of relationships, some embodied in the recorded law of the country and others as part of its traditions and conventions. Operating at this level, those of us that act in accordance with these precedents are considered to be well-behaved and those that break the conventions or laws are considered to be ill-behaved or even immoral.

We may feel that some of our present laws fall somewhat short of the ideal, but the intention of the lawmakers should be to make life easier and more measured for all of us, while maintaining our freedom of choice insofar as it is consistent with that aim. In general, in Britain there is a principle of law that requires us to desist from harmful actions, rather than requiring us to perform good actions. We may choose our own salvation, but must not deny this choice to others.

In the economic hierarchy illustrated, in the lower four levels — from government down to production — the details of relationships differ from state to state, and from time to time. The professions usually operate at level 4 on the diagram, although their influence permeates all levels. In their international or transnational roles they may be called upon to base their decisions upon higher considerations of humanity and justice, and upon the quality of the environment as a whole.

On crossing the boundaries of particular nations, we encounter different traditions of law and order, some of which will be in conflict with our own national statutes. These conflicts may lead to doubts and confusion as to the nature of right and wrong and may lead to mistrust and misunderstanding, if not to physical conflict.

> *A contractor — to whom you have just awarded a major contract — offers to fly you and your wife to an interesting conference in Cannes this Spring. Do you accept?*
> No!

Beyond the limits of the nation state and its laws and sanctions, lie humanity and justice. They relate to the well-being and values of mankind as a whole. Here we can begin to understand man and his relationship to the universe. We can begin to consider the universal qualities of justice and mercy. Most of the values reflected in the world's great teachings and traditions are expressed at this level, although they may or may not be incorporated in the laws and traditions of a particular nation at a particular time in history.

At the level of understanding and wisdom, we shall find educated men whose understanding is based upon the teachings of the wise. They may not have such wisdom themselves, but the precepts upon which they base their judgements and opinions will be good. The quality of their work is such that it can initiate and sustain certain attitudes towards law and tradition in a society for many centuries. Clearly, the responsibilities of such people are great and only those with the highest integrity can be entrusted with such authority.

> *If you govern the people by laws, and keep them in order by penalties, they will avoid the penalties, yet lose their sense of shame. But if you govern them by your moral excellence, and keep them in order by your dutiful conduct, they will retain their sense of shame, and also live up to this standard.*
> Confucius

If a great nation accepts its precepts from lesser men, then misery and suffering must follow. Examples of this have been seen in the recent histories of Germany, Russia and other countries. This is one of the dangers of a wholly secular society, deriving its values from the necessarily limited understanding of one individual unsupported by the traditions of more mature cultures. The traditions of wisdom and learning usually acknowledged as wise and just are commonly presented to the lower orders of the social hierarchy by prophets and teachers who have realised something of what has been called the ultimate good.

2.3.4. The consequences of our decisions

The third dimension of ethical behaviour is concerned with the consequences as well as the qualities of our actions, their range and longevity (see Fig. 2.4). Our behaviour is judged by society by its consequences, by our reliability and responsibility, and by our trustworthiness. If our actions are inconsequential, or only affect our own quality of life, then we can be ignored The consequences of our actions can be limited to ourselves or our immediate family, but are more likely to affect our society. They can affect mankind or even creation as a whole. Major engineering projects can affect the quality of life of many people for decades, if not centuries — indeed, it is often intended that they should. But they can also cause major disasters.

> *An irrigation scheme in the north of Sri Lanka was first built in about 200 AD. No doubt it displaced several small communities during its construction. It has served and continues to serve the community well. How do we assess the merits of the rival goods involved in the planning and construction of such a project?*
> By reference to the hierarchy of values given below.

Recognising that actions varied greatly in their significance, the ancient Vedantic teachings of India postulated five levels of discrimination or decision making, according to their consequences, ranging from the individual to the universal. These were considered in terms of the effects of actions, identified by reference to the following questions.

- What is good for me?
- What is good for my family?
- What is good for this society?
- What is good for mankind as a whole?
- What is good for creation?

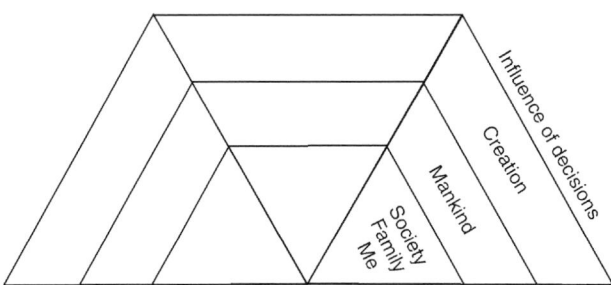

Figure 2.4

It is clear that at each level, the decisions made affect an increasingly large number of people and to an ever-widening extent, creation as a whole. It is important that we appreciate the range of influence of our decisions and therefore the range of understanding and objectivity required if these decisions are to be 'good'.

I am offered an Indian take-away for supper:
Do I say *'No thank you, I prefer a Chinese'*?
Do I say *'Yes please'*, because I know that the rest of the family will enjoy it?
Do I say *'No'*, because the particular supplier of ingredients is exploiting child labour?
Do I say *'Yes'*, because the trade generated will help to improve the Global Economy?
Do I say *'No'*, because the excessive use of water is drying out the aquifer, causing desertification and increasing global warming?

As professionals we are more likely to be concerned with decision making in the upper levels shown in Fig. 2.4, although our personal or family wishes may impinge upon our decisions from time to time. Our professional responsibilities require us to act as our client's agent, and also to consider the impact of our decisions upon those whose quality of life is likely to be affected. Generally, we can organise our lives to avoid or diminish such conflicts of interest, but if this is not possible — perhaps in some emergency situation — then the consideration of the higher level needs would normally take precedence over the lower.

It is your daughter's fifth birthday, you are to be the star entertainer this afternoon. This morning a mobile crane overturned on site, causing extensive damage and trapping the operator in his cab. The emergency services are dealing with the resultant problems, but you are the senior staff member on duty. Do you leave it to the fire brigade to resolve the crisis?
Your daughter must learn that there are some responsibilities that are more demanding than her party!

Potential conflicts of conscience, law, professional position and personal expediency are quite common. As indicated earlier, the resolution of ethical problems is rarely susceptible to logical analysis. Judgement is required. This involves an iterative, networking approach to conflict resolution and determination. Judgements frequently emerge out of the gathering and appraisal of the information surrounding the particular situation. Wherever possible the nature of a project should be considered relative to all these levels of discrimination.

2.3.5. The interactive ethos — our world

We can now begin to bring together the three sections of our model of the nature and context of ethical behaviour (see Fig. 2.5). These are

- the nature of the individuals and of the factors that have formed their personal value systems
- the nature of the societies in which the project is being executed and of their particular standards of acceptable behaviour — in international projects, or projects being designed for other nations or cultures, this is particularly important
- the consequences and quality of our actions.

All these are relevant to the nature and work of the engineer. In Chapter 3 we will explore the particular nature of professional actions, and the organisations of professionals within societies, together with their formulations of acceptable, or mandatory, standards of ethical behaviour for their members.

It is of value to note that the work of the engineering professional involves decision taking beyond the area of the limited nation state, with its relatively clearly defined laws, regulations, customs and conventions. The area of operation has been indicated in Fig. 2.6, entering the areas of the acceptance of values as a matter of personal responsibility, requiring knowledge of humanity and justice beyond the reach of the law, and having very far-reaching effects — both in scale and in time.

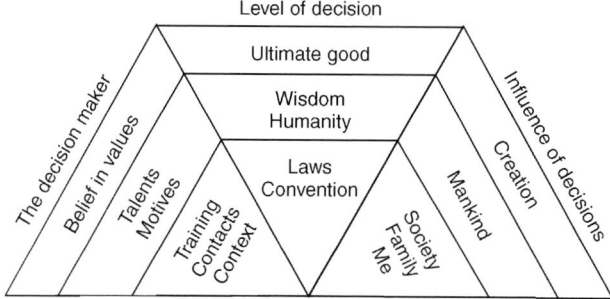

Figure 2.5

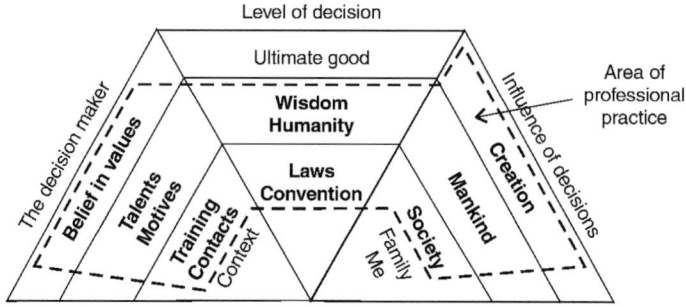

Figure 2.6

This challenge is a matter of professional satisfaction as well as of professional responsibility. The practical implications of such responsibilities will be seen in the ethical audit example set out in Appendix 1.

3

The professions

3.1. The nature of a profession

The Oxford Shorter Dictionary defines a profession as: 'A vocation, a calling, one requiring advanced knowledge or training in some branch of learning or science'.

To be a member of a profession, therefore, is to

- have specialised knowledge and skill
- have power — the power of knowledge and the capacity to affect society
- have autonomy of practice — this varies according to employment context
- have a monopoly or near monopoly of a particular skill
- have undergone an extensive period of training which includes not simply skills but a strong intellectual element
- be a member of a professional body which is responsible for regulating standards, protecting rights of practice and ensuring proper training.

> *Professionals have to have autonomy. They cannot be controlled, supervised or directed by the client. Decisions have to be entrusted to their knowledge and judgement. But it is the foundation of their autonomy, and indeed its rationale, that they see themselves as affected with the client's interest.*
> Peter F. Drucker

3.2. The professional engineer
Professional practitioners, of all disciplines, need six core qualities

- integrity — openness and honesty, both with themselves and with others
- independence — to be free of secondary interests with other parties
- impartiality — to be free of bias and unbalanced interests
- responsibility — the recognition and acceptance of personal commitment
- competence — a thorough knowledge of the work they undertake to do
- discretion — care with communications, trustworthiness.

These professional qualities, linked with the personal virtues discussed in Chapter 2, enable the professional to maintain principles and fulfil responsibilities, and to do this in a context which is often complex and unclear. With growing specialisation in the division of work there is the danger of the 'expert' becoming focused purely on technical matters. It is important to note how this can affect the perception of the professional engineer, discouraging holistic thinking and inhibiting the full breadth of his work.

Integrity
This involves the discovery and communication of the truth. It is not, however, simply truthfulness or the avoidance of lying, but the capacity to communicate the truth in such a way as to enable the client and others to make informed decisions. Honesty and integrity are essential for the development of trust.

Integrity leads to a concern for the whole situation in decision making, including an awareness of the professional's own attitudes, standards and value systems. It leads to consistency of character and operation in different situations and contexts. A good test of this is not just consistent operation in different contexts but also how the individual operates when there is nobody around, testing consistency and truth against personal attributes.

Integrity ensures that the professional does not accept 'moral distance' or the majority view of any particular group, but is true to his own objective understanding of the situation.

Independence
Independence is not so much a virtue as a state. The professional is independent of pressure from any interested group. To be involved with

either the employer or an action group in such a way that professional judgement is clouded, is to lose professional autonomy. To achieve this independence the professional must understand the situation and the key players in that situation. Independence enables the person not to be drawn into the concerns of any particular group. It presupposes the willingness and ability to take responsibility for decisions.

By virtue of his special knowledge the professional has an intrinsic authority. He also has an extrinsic authority vested in him by virtue of the position he holds. It is important to recognise that such authority leads many people to accept directions given by him despite any clearly unacceptable outcomes. Their personal responsibilities may be denied and transferred to the authority figure. Professionals must respect and encourage others to accept their own responsibilities.

Impartiality

Impartiality enables the professional to fulfil his contracts with his client, and to treat all parties equally. There is a need to avoid concerns for self-interest, for self-advancement and for the pressures of management roles, where priorities for the achievement of programme and cost targets may cloud professional judgements.

Responsibility

Responsibility involves the realistic assessment of skills and capacities and the acceptance of their possibilities and limitations. This is a core virtue in enabling the professional to acknowledge the responsibility of others, to accept personal responsibility and to work as a team — an essential part of modern engineering projects.

Problems arise when professionals develop an inflated sense of responsibility, seeing themselves as responsible for more than they can be or ought to be. Realism is critical to ensure the best contribution from all members of the team.

The engineer owes responsibility to

- the general public/society
- the users of the project
- the peer group of professional engineers and institutions
- clients and employers.

Engineers are responsible for

- their own actions
- duties accepted and implicit in their work and institutions
- legal requirements of the nations in which they practise

- legal obligations imposed by contract
- the greater moral consequences of their actions, and of those for whom they are responsible.

Most members of the construction industry are now described as 'professionals' and their roles have merged. Consultants are more concerned with the commercial aspects of their practice, with marketing and fee bidding strategies. Contractors are leading teams of 'professionals' in management contracts and in design-and-build agreements. Roles are further combined and interwoven by the addition of financial and legal specialists in Private Finance Initiative (PFI) projects — finance, design, build, operate and handover projects. These all have a considerable influence on the concepts of professional ethics and responsibilities.

But this interweaving of roles requires an even more conscious understanding of the responsibilities of professionals. It is difficult to recall the responsibilities as an agent rather than as a contractor during the very close relationships arising during a total procurement process. Professionals are still responsible for making decisions that others have not the skills and training to make. The need for integrity, independence, impartiality and responsibility is even greater. It is important to avoid a crude view of professionalism which results in a concentration on the detail of the technical brief, ignoring social factors and the wider implications of one's decisions. The responsibility for one's role in collaborative work within and beyond the profession is of prime importance.

Competence
Perseverance is necessary for the acquisition of technical competence and for its application in solving technical problems. This involves the capacity to strive for and maintain competence in professional practice. Competence is essential if the professional is to fulfil his or her responsibilities to society. The ethical requirement is in the acknowledgement of the need to be competent, rather than in the competence itself. Skills can be misused.

An essential element of professional practice is the capacity to reflect upon, to evaluate, and to accept different levels of competence, and to ensure that the necessary skill is deployed, even if it has to be procured elsewhere.

Discretion
In the course of projects the professional will become aware of many aspects of the affairs of clients, contractors or other interested parties,

and for him to retain the essential trust of those parties will be important. Information must be treated on a 'need to know' basis, and not transmitted unnecessarily to others. In some cases, on defence projects for example, consultants will be under an agreement not to divulge information to others. The responsibility placed upon the professional to act on behalf of others will be increased, his impartiality and independence becoming of great importance. He must be able to raise matters of concern with his client and, if need be, with higher authorities if he is seriously concerned about the consequences of some action. The professional does not, in his professional role, become a servant.

3.3. The role of the engineer

> *The Engineer is a Mediator between the Philosopher and the Working Mechanic, and like an interpreter between two foreigners, must understand the language of both, hence the absolute necessity of possessing both practical and theoretical knowledge.*
> Henry Palmer. Inaugural address at the Institution of Civil Engineers, 2 January 1818

If this definition of the engineer is accepted then the professional engineer, acting as such, can never truly be an employee. The true professional needs and retains the freedom to act in accordance with personal judgement, unbiased by the overriding needs of his employer.

The engineer can be considered as playing several roles in society. He can be

- saviour — the engineer as the key player in the creation of a Utopian society with the development of technology and material prosperity for all
- guardian — ensuring the best interests of society, based upon engineering knowledge
- bureaucratic servant — a servant to managers, translating directives into achievements
- social servant — an obedient social servant to others
- game player — playing to political and economic game rules — playing successfully within the organisation — competing to win.[1]

In all of these roles the engineer can act in different capacities, he can be a consultant, an employee, a manager, and employer. These roles can frequently produce conflicts of interest.

---

[1]Martin M.W. and Schinzinger R. *Ethics in engineering*. McGraw Hill, New York, 1989.

When the engineer becomes a manager a further complication is added. The most important conflict is that involved in the decision making processes. This springs at heart from a conflict of loyalties. For example, as a manager he may be held to have an overriding loyalty to the company, to its efficiency and profitability. As professionals, engineers have responsibilities for the wider interests of safety, health and welfare of the public. When these criteria are applied to 'choices', conflict is inevitable, and the need for professional qualities becomes very important. These qualities, together with a knowledge of the consequences of his actions, and of the nature of the society in which the actions are taking place, are essential if the engineer's decisions are to be clear and independent, and truly professional.

A clear case where conflicts can arise is in the negotiations of contractual claims for additional work or for delays. Clear procedures are usually available. The professional engineer has a responsibility for fairness. Clear statements of the circumstances are essential, and the interests and responsibilities of all parties should be acknowledged. The guiding principle is one of equity, not of maximising gains.

Engineering projects, particularly civil engineering projects, are often multifaceted endeavours. They can be complex, unique, built to strict deadlines of cost, time and quality, of high value and high in uncertainty. They are invariably the result of team efforts, often involving clients inexperienced in engineering. These constraints require detailed coordination and project management, a role often undertaken by the professional engineer

> *The function of the project manager is to foresee or predict as many of the dangers and problems as possible and to plan, organise and control activities so that the project is completed successfully in spite of the risks.*
> Locke — 1992

The engineer's technical skills must be augmented by other skills involving human relationships, administration, information management, programming, financial reporting and control, and quality management. Managers are not usually considered as professionals, being concerned with matters and standards that are more organisation centred, whilst engineers are governed by standards beyond the organisation, usually set by their training, codes of practice and institutional standards.

The professional engineer may be considered as serving society. This gives him a particular social status, demanding autonomy of decision making to ensure that his judgement is not compromised by matters of expediency or finance. The engineer can be seen as a custodian of the concerns of many different groups of individuals or interests, including the environment and those who have no present voice, such as future generations. The engineer has to survive, however, and the engineering company needs to be profitable in order to stay in business. At the same time the engineer cannot operate simply in a market-orientated way. The creative work of the engineer cannot be seen as simply a product. It involves a great deal of skill unavailable to the layperson. The professional engineer is not simply selling his services in the marketplace, with the buyer being responsible for deciding between different products (*caveat emptor* — let the buyer beware). He is rather offering his services, with the buyer needing to trust his (the engineer's) skills and judgement (*credat emptor* — let the buyer trust).

3.4. The professional institution

> *A Society for the general advancement of Mechanical Science and more particularly for promoting the acquisition of that species of knowledge which constitutes the profession of a Civil Engineer, being the art of directing the Great Sources of Power in Nature for the use and convenience of man . . .*
> Charter of the Institution of Civil Engineers, 1828

The virtue of integrity is not simply an individual value, but one which applies also to the professional institution and management groups.

The profession as a whole has to be consistent and be able to relate values to practice. Without this the recognition of the profession as a body of experts concerned for public welfare will be eroded and with that will go the trust essential to the functioning of any professional relationship.

The existence of an institution of professionals is essential for the development and maintenance of professional virtues. It enables the individual professional to reflect upon his own integrity and to learn from the experience of others, transmitting the culture from generation to generation. It provides an external perspective which enables proper reflection and responsibility.

An Institution can

- enable the professional development of moral awareness, skills, responsibility and identity, through codes, dialogues and training

- ensure that its processes and organisation are conducive to the development of moral responsibility
- provide support and the opportunity for professionals to work through decision making and any conflicts of interest
- regulate the practice of the individual professional
- play a major role in communicating with the public
- set standards for admissions to institutions and for initial and continuing professional training
- act as a learned society, contributing to the advancement of science and technology of engineering.

The professional institution must avoid the moral ambiguity of acting as a body protecting the interests of its individual members such as a trade union or trade association. It must be concerned with ensuring the highest standards of professional service to the community by regulating standards and criteria for membership. The institution itself has to develop a means of reflecting on its own ethos and ethical culture. This is a prime responsibility of the council of an institution, requiring regular reviews of the professional code and ensuring sound practice by all its members.

3.5. The professional code

> *All these trust in their hands; and everyone is wise in his work.*
> *Without these cannot a city be inhabited; and they shall not dwell*
> *where they will . . . . . But they will maintain the state of the world*
> *and all their desire is in the work of their craft.*
> Ecclesiasticus — Chap 38 v31/32 34

3.5.1. The nature of codes

Professional codes enable ethical reflection and development, and are a means of developing the integrity of the professional body. The public declaration of the principles and views on responsibility and practice of a professional organisation provide a benchmark against which the practice of the profession can be evaluated.

Ethical behaviour is an essential component in effective human societies. The aim is to enable collaboration between individual members, thereby improving the overall quality of life and fulfilment in the society, by creating an atmosphere of trust. The increasing awareness of social, international and global issues has expanded our awareness and is moving towards a sense of responsibility which extends far more widely than was the case when professional standards were first formulated. The Institution of Civil Engineers (ICE), therefore, at the instigation of the late President, Edmund Hambly, conducted a review of its Rules of Practice in 1996/97 and accepted a redrafted code in 1998.

The changing context in which professionals practice required a review of traditional ethical practices and requirements. There was a need in many areas to enhance professional standards and to increase awareness of the changing perceptions of the community at large as to the nature of professional integrity and practices. Environmental and human issues, the interests of minority groups, international competition, and the growth of information technology (IT), all affect our activities.

The recent redrafting of the Rules of Practice of the ICE have considered not only the interests of clients and fellow professionals, but have further extended the range of interests to consider the responsibilities of Engineers to the environment and to all who are affected by their work. As set out in chapter 2, as part of the 'Decision Maker's World', the levels of responsibilities of all human beings, not only of engineers, can be formulated as ranging from individual 'goods', through family care and the interests of various social groups, such as professional organisations, to mankind as a whole and the general quality of our habitat, which is, these days, almost universal. These can be summarised as:
What is good for

- the environment
- mankind
- my associates (other professionals, citizens, countrymen)
- my family
- me.

Many apparent conflicts of interest can be resolved by reference to the next level 'up' in this hierarchy of values. On the assumption that members are individually responsible for their own and their families' welfare, the Rules of Practice are concerned primarily with the upper three categories, which are as follows.

General issues: Responsibilities for the effects of our work on the global environment. This category would include not only major issues of 'global warming', energy conservation, land use, etc., but also economic issues, such as employment, sources of materials, etc.

These aspects are covered by Rules 1,2 and 3

Societal issues: Responsibilities for the quality of life of other human beings. This category would cover disturbance to local populations and all business transactions.

These aspects are covered by Rules 4, 5, 11, 12 and 13

Professional issues: Responsibilities to our fellow professionals. This category would include upholding the profession's status and ensuring competence and the appropriate transfer of skills from generation to generation.

These aspects are covered by Rules 6, 7, 8, 9, 10, 14, 15 and 16

3.5.2. Comparative examination of ethical codes

In reviewing the ICE Rule, a synthesis of the codes of the ICE, ASCE, EC, Civil Service, WFEO, FIDIC and the Nolan report on Standards in Public Service was prepared. Summaries of the main features of these codes are given in Tables 3.1 and 3.2.

The key factors which are evident in all the codes considered are set out below.

- Due regard must be given to health and safety, both immediate and in the long term.
- The need to enhance, or to diminish adverse effects upon, the quality of the environment — systematic reviews of all aspects of impact, including cost-benefit analyses, should be routine on all projects.
- The importance of ensuring public understanding of all the parameters of project development, and to involve them whenever practicable in the decision making process. The conflicts of interest that arise because of the hierarchy of ethical factors — from the personal to national or even international — need to be clearly expressed and understood by both the professionals involved and by the public.

A knowledge of the wide-ranging consequences of major engineering works is reflected in the increasing call for full environmental impact assessments for all major projects in the European Economic Community (EEC). The requirements of the EEC directive on environmental impact assessment were also considered and are summarised in Table 3.3.

Engineering responsibilities for the environment are currently governed by laws controlling environmental matters. Stringent regulations now exist at local, national and international levels, which are designed to ensure that the industry conducts its business in an environmentally friendly fashion. British Standard 7750 — *Specification for environmental*

Table 3.1 (pages 39–41). Comparison of various professional codes of ethics

Organisation	Code of ethics
Institution of Civil Engineers	Regard for public interest, health, safety, Integrity Fidelity to employers Not to injure professional reputation Not to canvass or offer commissions No excessive publicity Avoid conflicts of interest, no receipt of gratuity Fair competition Relate to local codes or follow own in other countries Criminal offences are evidence of improper conduct No supplanting of appointed engineers Not to act as medium for payments from clients Further education of others
Engineering Council	Enhance the quality of the environment Adopt balanced, disciplined, comprehensive approach Make systematic reviews on environmental issues Encourage management to adopt positive environmental policies Follow professional codes Comply with and understand Law Maintain competence through CPD Encourage public understanding
UK Civil Service	To serve elected government with integrity, honesty, impartiality Civil servants owe loyalty to elected government Discharge public functions reasonably and lawfully Obey laws, international laws, not to imperil justice Obey professional codes Make all information available to minister Not to mislead or deceive ministers Work efficiently Not to use position to further private interests Maintain dignity of position and trustworthiness Not to misuse information or disclose confidential material To follow laid down procedures for reporting irregularities in instructions If still in doubt report matter to Civil Service Commissioners Not to decline authorised instructions — or to resign

Table 3.1 continued

Organisation	Code of ethics
American Society of Civil Engineers	*Fundamental Principles:* Use knowledge and skill to enhance human welfare Be honest and impartial in serving public, employers, clients Increase competence and prestige of profession Support work of professional societies *Fundamental Canons:* Hold paramount, safety, health, welfare of public Perform services only in areas of own competence Make public statements only in objective/truthful manner Act as faithful agents, avoid conflicts of interest Base reputation on merit, not on unfair competition Uphold dignity of profession Continue to maintain own competence and that of others.
World Federation of Engineering Organisations	Hold paramount, safety, health, welfare of public Perform services only in areas of own competence Act as faithful agents, avoid conflicts of interest Continue to maintain own competence and that of others Deal fairly and with good faith towards all parties, give due credit and honest criticism Communicate awareness of environmental and social consequences objectively Present possible consequences of disregarding engineering judgements Report any illegal or unethical engineering decisions or practices
FIDIC	*Responsibility to Society:* Accept the responsibility of the consulting industry to society Seek solutions compatible with sustainable development Uphold dignity and reputation of the consulting industry *Competence:* Maintain knowledge and skills — Apply due care and attention to service rendered Perform services only when competent so to do

Table 3.1 continued

Organisation	Code of ethics
FIDIC	*Integrity:*
	Provide all services with integrity
	Impartiality:
	Be impartial in giving advice and making judgements or decisions
	Inform clients of conflicts of interest
	Do not accept remuneration prejudicing independent judgements
	Fairness to others:
	Promote 'qualifications-based selection'
	Do not injure reputation or business of others
	Do not supplant previously appointed consultants
	Do not accept the work of others without confirmation by client of termination of their appointment
	In reviewing the work of others behave with courtesy and consideration
	Corruption:
	Do not accept or offer remuneration to influence appointment of consultants or clients, or that may affect consultant's impartiality
	Cooperate with any legitimate investigation into any contract administration

management, ISO 14001 — *Environmental management systems: specification with guidance for use, Environmental assessment directive 85/337/EC* all provide guidance and control to ensure that projects receive an appropriate level of environmental assessment and analysis.

The main areas of interest identified in the several codes were divided into three classes, those of concern to the environment and mankind generally, those of concern to particular societies, and those of interest to the professions themselves.

The five levels of consequence listed in Section 2.3.4, together with the criteria of impact assessment, illustrate the effects of the major projects upon the community and the environment, and provide some guidance in the preparation of a revised code under the following three headings.

Table 3.2 Committee on Standards in Public Service (Lord Nolan). The Seven Principles of Public Life

---

Selflessness
Holders of public office should take decisions solely in terms of the public interest. They should not do so in order to gain financial or other material benefits for themselves, their family, or their friends.

Integrity
Holders of public office should not place themselves under any financial or other obligation to outside individuals or organisations that might influence them in the performance of their official duties.

Objectivity
In carrying out public business, including making public appointments, awarding contracts, or recommending individuals for rewards and benefits, holders of public office should make choices on merit.

Accountability
Holders of public office are accountable for their decisions and actions to the public and must submit themselves to whatever scrutiny is appropriate to their office.

Openness
Holders of public office should be as open as possible about all the decisions and actions they take. They should give reasons for their decisions and restrict information only when the wider public interest clearly demands.

Honesty
Holders of public office have a duty to declare any private interests relating to their public duties and to take steps to resolve any conflicts arising in a way that protects the public interest.

Leadership
Holders of public office should promote and support these principles by leadership and example.

These principles apply to all aspects of public life. The committee has set them out here for the benefit of all who serve the public in any way.

3.5.3. The public interest

Environment
A great deal has been said about the crucial responsibilities of the civil engineer with respect to the environment, much of it slanted towards the ecological and topographical impact of projects. It is important to

Table 3.3. EEC Environmental Impact Directive (85/337/EC)— ethical code

Basis of assessment	
The existing characteristics of the location	Land, water, landscape Climate Ecology Population Services
The likely impact during construction	Emergencies Waste products Noise Workforce Services requirements Transportation Material requirements
The impact of the completed development	Emergencies Waste products Noise Population changes Services requirements Transportation Raw materials Company expenditure Site utilisation

The structuring of the environmental report	Description of project
	Impact of project:
	    adverse effects
	    beneficial effects
	    duration effects
	    reversibility
	    significance local/strategic

maintain a balanced view of the overall needs of societies, to consider economic, social and political interests as well as those of the physical environment.

The key qualities required of the professional in these general areas are concern for the public interest and evident integrity. Public confidence in the integrity of the professions is essential to the provision of a responsible service. To this end professional practitioners must not only be competent and trustworthy, but must be seen to be so.

The key factors concerning integrity and the public interest, from the point of view of the drafting of codes of conduct, are set out below.

- The adoption of a balanced, disciplined and comprehensive approach to problem solving. It should be clear to all that professional judgement and skill are being used to enhance human welfare and diminish negative effects of developments. Working practices should be efficient and economical.
- The discharge of public functions reasonably and lawfully. Conflicts of interest must be avoided, advice must be independent and impartial, free from any personal interests of the professional.
- As far as possible all information should be freely available. It should not be misused to mislead or deceive, nor should confidential material be disclosed. Public statements must be made only in an objective and truthful manner.
- Procedures should be established and followed for reporting irregularities and conflicts of interest. Requirements by clients for the professional to accept instructions which are perceived to be against the public interest must not be accepted.

3.5.4. Societal interests

The relationships between the members of a particular society or nation are usually contained within the laws and convention of that society. The qualities of independence, impartiality and responsibility discussed earlier are particularly relevant here. The conduct of the professional requires a self-regulation which is greater than the regulations prescribed by the legal restrictions and requirements. The definition of the degree and nature of such regulation is embodied in the professional codes, but only in very general terms, leaving the professional to use his or her personal qualities to make appropriate responses to particular situations. It is neither possible nor advisable to be prescriptive about all the details of professional practice.

All transactions, however, should be in accordance with the laws of the nation state, as well as being morally responsible. To this end, codes of behaviour require

- that professionals understand and comply with the laws of the communities within which they are practising, and with international law

- that where professional codes exist, these should also be followed
- that where neither laws nor codes exist then principles of good practice should be established between all parties — clients, consultants, contractors, suppliers, etc.

3.5.5. Professional interests

The demonstration of the reliability and trustworthiness of the members of a profession requires competence, personal and professional dignity, and respect and consideration for fellow professionals.

Technical competence should be rigorously assured. The training and testing procedures of the professional institutions should be made known to the public, and their value clearly expressed.

Whilst technical competence alone cannot guarantee trustworthy professional service, it is an essential factor that must be present. To this end the professional person must

- be fully trained and broadly educated, with an understanding of the context within which decisions are made, as well as of their technical soundness
- ensure that he/she maintains competence by continually updating and extending the understanding of technical and professional developments
- contribute to the education and training of others, both during the formation period and throughout their professional development.

The professional engineer has a personal duty towards fellow professionals to

- carry out systematic reviews of personal ethical standards and to understand the significance of the ethical dimensions of means as well as ends
- uphold the dignity, respect and trustworthiness of the profession
- adhere to established professional codes of practice
- not use his position to further private interests
- perform services only in areas of his own competence
- ensure that reputations are established on merit, not on unfair competition
- support the work of professional societies.

In business dealings with fellow professionals the engineer must not

- injure the professional reputation of others
- supplant appointed engineers

- canvass excessively or offer commissions
- benefit from unfair competition.

All these factors need to be incorporated within professional codes simply and comprehensively.

3.6. The Revised Code of Practice

3.6.1. The previous Rules

The previous 14 rules of the Institution of Civil Engineers could be summarised briefly as follows.

1. Regard for public interest, health, safety
2. Integrity
3. Fidelity to employers
4. Not to injure professional reputation
5. Not to canvass or offer commissions
6. No excessive publicity
7. Avoid conflicts of interest, no receipt of gratuity
8. Fair competition
9. Relate to local codes or follow own in other countries
10. Criminal offences are evidence of improper conduct
11. No supplanting of appointed engineers
12. Not to act as medium for payments from clients
13. Further education of others
14. Assist with continuing professional development.

3.6.2. The revised Rules of Practice

All professional codes of practice need continuous review to ensure compatibility with contemporary factors, changing work patterns and changing legislation. The *Revised Rules of the Institution of Civil Engineers* (1999) are set out in Table 3.4. As stated above, personal and family concerns of members are not included, since these are common to all individuals, and are not subject to the regulations of a professional body, except so far as they may affect a person's dignity or standing in society and thus impact upon the trustworthiness of the profession.

Comments on each of these rules are given below.

Rule 1. A member shall discharge his professional responsibilities with integrity and shall not undertake work in areas in which the member is not competent to practise.

Comment

Integrity implies the adoption of a balanced, disciplined, and comprehensive approach to problem solving. Members shall, when called upon to do so, demonstrate that professional judgement and competence are being applied in their work as agents of the employing authorities or clients.

Rule 2. A member shall uphold the dignity respect and trustworthiness of the profession at all times.

Comment

Professional engineers are expected by the public and by Clients and Employers to give impartial and competent advice. They are in effect 'licensed' by their membership of a distinguished qualifying Institution, and the confidence of the public rests on the standing of that Institution, whose integrity must be maintained and demonstrated by the behaviour of its members.

Rule 3. A member shall have full regard for the public interest, particularly in relation to the environment and to matters of health and safety.

Comment

Members should ensure that systematic reviews of all aspects of a project's impact upon the environment, including the justification of the need for the project, and economic, social and political factors, and cost benefit analyses, should be undertaken to diminish any adverse effects. Disadvantageous cost benefit analyses may be acceptable if social gains are obtained. Due regard must be given to health and safety, both immediate and in the long term. As far as possible all information should be freely available. Public confidence is essential to the provision of a responsible service. Public statements shall be made in an objective and truthful manner.

It is particularly important to ensure public understanding of all the parameters of project development, and to involve the public whenever practicable in the decision making process. The conflicts of interest that arise because of the hierarchy of ethical factors — from the personal to national or even international — need to be clearly expressed and understood by both the professionals involved and by the public.

Where confidentiality is required in the Engineer's agreement, perhaps for defence, security or commercial reasons, members should recognise their responsibilities for public and third party interests.

Rule 4. A member must understand and comply with the Laws of the communities within which he practises and with International Law.

Where Professional Codes exist in the country concerned, these should be followed. Where neither Laws nor Codes exist then the Institution's Rules of Professional Conduct should be followed.

Comment
All transactions should be within the Law and follow local conditions of professional practice in where these exist. Members are morally responsible for their activities.

Rule 5. A member, without disclosing the fact to the Employer in writing, shall not be a director of, nor have a substantial interest in, nor be an agent for any company, firm or person carrying out any contracting, consulting or manufacturing business which is, or may be involved in the work to which the employment relates; nor shall the member receive directly or indirectly any royalty, gratuity or commission on any article or process used in or for the purposes of the work in respect of which the Member is employed unless or until such royalty, gratuity or commission has been authorised in writing by the employer.

Comment
Conflicts of interest must be avoided. An engineer must be in a position to ensure that advice is independent and impartial and free from any personal interest. The engineer should be faithful to the Public and his Client's needs and discharge his duties with impartiality.

Rule 6. A member shall afford such assistance as he may reasonably be able to give to further the Education and Training of candidates for the profession.

Rule 7. A member, either individually or through the member's organisation as an employer, shall afford such assistance as may be necessary to further the continuing development of the individual and of other members and prospective members of the profession in accordance with the recommendations made by the Council from time to time.

Comment
A professional person must be broadly educated and fully trained, with an understanding of the context within which decisions are made, as well as assuring the technical soundness. Developments in technology place upon engineers a duty to maintain competence throughout their careers.

Rule 8. A member shall not maliciously or recklessly injure or attempt to injure, whether directly or indirectly, the professional reputation, prospects or business of another person.

Comment
In business dealings with fellow professionals the engineer must not injure the professional reputation of others, supplant appointed engineers, canvass excessively or offer commissions or benefit from unfair competition.

Rule 9. A member who shall be convicted by a competent tribunal of a criminal offence which in the opinion of the disciplinary body renders him unfit to be a member shall be deemed to have been guilty of improper conduct.

Comment
Engineers known to be guilty of illegal actions betray the trust due of professional practitioners, and are liable to have their membership withdrawn.

Rule 10. A member shall not, in any manner derogatory to the dignity of the profession, advertise or write articles for publication, nor shall authority be given for such advertisements to be written or published by any other person.

Comment
The maintenance of professional dignity by members is of paramount importance in maintaining the confidence of the public.

Rule 11. A member shall, consistent with safety and other aspects of the public interest, endeavour to deliver to the employer or client cost effective solutions. A member shall not comply with any instruction requiring dishonest action or the disregard of established norms of safety in design and construction.

Comment
The engineer is the client's agent, and as such must serve his needs faithfully and diligently, but this does not take precedence over the wider needs of society.

Rule 12. A member shall discharge duties to the Employer and the Client both impartially and with complete fidelity. A member shall not receive any benefit or advantage relating to the work unless authorised by the Employer or Client in writing. A member shall also declare any pecuniary interest in any other organisation involved with the work that is being undertaken for the Employer and the Client. Such a declaration may be made at the outset of the engagement or at any time when such a situation may become apparent during the progressing of the work for the Employer or Client.

Comment
Conflicts of interest must be avoided. An engineer must be in a position to ensure that advice is independent and free from any personal interest. The engineer should be faithful to the public and his client's needs and discharge his duties with impartiality.

Rule 13. A member shall not be the medium of payments on the Client's behalf unless so instructed by the Client or Employer; nor shall the member in connection with work on which employed, place contracts or orders except with the authority of and on behalf of the Client or of the Employer as appropriate.

Comment
Where members are engaged as sub-consultants and on design/build and/or finance contracts they should satisfy themselves that the terms of the main or superior contract do not commit them to actions which require a contravention of any of these rules of conduct.

Rule 14. A member shall not improperly canvass or solicit professional employment nor offer to make by way of commission or otherwise payment for the introduction of such employment unless disclosed to the employer.

Rule 15. A member shall not, directly or indirectly, attempt to supplant another Engineer.

Rule 16. If requested to comment on another engineer's work a member shall act with integrity, and, except for routine or statutory checks or when the member's Client or Employer requires confidentiality, the member should advise the other engineer of his involvement.

Comment on 14, 15, 16
In business dealings with fellow professionals the engineer must not injure the professional reputation of others, supplant appointed engineers, canvass excessively or offer commissions or benefit from unfair competition.

Table 3.4. The Institution of Civil Engineers — Summary of Rules of Conduct

1	A member shall discharge his professional responsibilities with integrity and shall not undertake work in areas in which the member is not competent to practise.
2	A member shall uphold the dignity, respect and trustworthiness of the profession at all times.
3	A member shall have full regard for the public interest, particularly in relation to the environment and to matters of health and safety.
4	A member must understand and comply with the Laws of the communities within which he practises and with International Law. Where National professional Codes exist, these should be followed. Where neither Laws nor Codes exist then the Institution's Rules of Professional Conduct should be followed.
5	A member, without disclosing the fact to the Employer in writing, shall not be a director of, nor have a substantial interest in, nor be an agent for any company, firm or person carrying out any contracting, consulting or manufacturing business which is, or may be involved in the work to which the employment relates; nor shall the member receive directly or indirectly any royalty, gratuity or commission on any article or process used in or for the purposes of the work in respect of which the Member is employed unless or until such royalty, gratuity or commission has been authorised in writing by the employer.
6	A member shall afford such assistance as may be reasonably given to further the Education and Training of candidates for the profession.
7	A member, either individually or through the member's organisation as an employer, shall afford such assistance as may be necessary, to further the continuing professional development of the individual and of other members and prospective members of the profession in accordance with the recommendations of the Council from time to time.
8	A member shall not maliciously or recklessly injure or attempt to injure, whether directly or indirectly, the professional reputation, prospects or business of another engineer.
9	A member who shall be convicted by a competent tribunal of a criminal offence which in the opinion of the disciplinary body renders him unfit to be a member shall be deemed to have been guilty of improper conduct.

Table 3.4. continued

10	A member shall not, in any manner derogatory to the dignity of the profession, advertise or write articles for publication, nor shall authority be given for such advertisements to be written or published by another person.
11	A member shall, consistent with safety and other aspects of the public interest, endeavour to deliver to the employer or client cost-effective solutions. A member shall not comply with any instruction requiring dishonest action or the disregard of established norms of safety in design and construction.
12	A member shall discharge duties to the Employer and the Client both impartially and with complete fidelity. A member shall not receive any benefit or advantage relating to the work unless authorised by the Employer or Client in writing. A member shall also declare any pecuniary interest in any other organisation involved with the work that is being undertaken for the Employer and the Client. Such a declaration may be made at the outset of the engagement or at any time when such a situation may become apparent during the progressing of the work for the Employer or Client.
13	A member shall not be the medium of payments made on the Client's behalf unless so instructed by the Client or Employer; nor shall the member, in connection with work on which employed, place contracts or orders except with the authority of and on behalf of his Client or of the Employer as appropriate.
14	A member shall not improperly canvass or solicit professional employment nor offer to make by way of commission or otherwise payment for the introduction of such employment unless disclosed to the employer.
15	A member shall not, directly or indirectly, attempt to supplant another Engineer.
16	If requested to comment on another engineer's work a member shall act with integrity, and, except for routine or statutory checks or when the member's Client or Employer requires confidentiality, the member should advise the other engineer of his involvement.

4

Major ethical issues

4.1. Introduction

The recognition of ethical dimensions in all aspects of professional engineering leads to proactive decision making and the reduction of potential ethical conflicts. Not all ethical issues can, however, be planned for, and the engineer can be faced with major dilemmas which have to be worked through 'from cold' (see ICE Rules 1,2,3,4,5,6,7,8,11,14 and 15).

In this chapter, the issues surrounding the more dramatic, reactive ethical problems, which may raise what seem to be intractable competing demands, are examined. Sometimes they will be solved through compromise, sometimes through taking a stand.

4.2. Reporting unethical issues

The need to report to a higher authority, beyond one's immediate superiors, may not be a common problem, but it can present itself as a major crisis, with strong emotional overtones. It is sometimes known as 'whistle blowing' — where the engineer considers that the fulfilment of his social obligations is more important than that to his company or colleagues. The decision to report demands careful judgement about the nature of the moral problems involved and about the consequences of such a decision.

The following cases illustrate some of the difficulties with which an engineer may be faced.

- A junior engineer on a building site is given the task of developing plans, which should be checked by a senior. It soon becomes apparent that these are not being checked but simply stamped and sent on for work. He mentions this to his immediate superior who dismisses the protest, asking how he thinks they are to get any building done if they follow all the rules to the letter. Where should he go from here?
- A civil engineer frequently walks past a building site for which he has no responsibility and over time notes several problems with safety and with the quality of work. What should he do?
- A senior consultant engineer who has followed the correct procedures in the development of a major city centre building has it pointed out to him by one of the undergraduates with whom he works in a university link, that there are serious design faults in the foundations of the building. What should he do?

These few cases, all based on real events, demonstrate some of the critical issues involved. The need to report may involve criticism of the actions of colleagues, of an organisation, or even of oneself. Whose responsibility is it to report in these situations? Should it only be the professional who is directly involved? Are there any alternatives to such reporting? What will the effects be on the professional, his family and career, on the organisation, etc? If it is right to report in this way, to whom should the professional report — to his professional body, to a government or other official body, to the senior management in his company, to concerned pressure groups, to the press?

The principal features of such unusual reporting are set out below.

- The information is conveyed outside approved organisational channels, or in situations where the person conveying it is under pressure from supervisors or others not to report it.
- The information is new to, or is not fully known by, the person or group to whom it is being sent.
- The information is believed to represent a significant moral problem concerning the organisation. Examples of such significant problems would be: criminal behaviour; unethical practices; injustices to workers within the organisation; threats to public safety; breaches of local customs.

The method of reporting may depend upon the potential effects of the action on the organisation, the professional, the public and any other relevant body. Reporting requires a moral decision, an estimate of the extent of the problem, the nature of the professional responsibility and the probable consequences of the various options. The reporter must be prepared to give way if his judgement is shown to be incorrect.

The professional requirements for the engineer to discharge his responsibilities with integrity, and to have regard for the public interest (ICE Rules 1, 3), are seen as of a higher order than his obligations to his client or colleagues.

It has been argued that there are two separate areas of decision — engineering and management — and that these should not be confused. The engineering decision is dominated by technical matters and by ethical concerns for safety. The management decision is different, focusing on responsibilities to the work force, the shareholders, the local community, the customer, and thus upon economic, legal and public image matters. Apart from those frequent situations where the roles of manager and engineer coincide, the engineer cannot avoid his responsibilities by delegating them to a manager. All major projects involve teamwork, and collective responsibility is more often the case than individual responsibility. His particular expertise may place the engineer in a better position to understand the risks involved. In such a case he has a duty to inform his colleagues that the risk or hazard exists, and should his colleagues ignore such advice then, he has a clear duty to report the facts to some other effective authority.

The ties of loyalty are said to be crucial to the effective and efficient operation of the economy, and extensive reporting beyond contractual relationships is said to have, ultimately, a detrimental effect upon the public good. A critical part of that efficient functioning is confidentiality. Such an argument builds up an adversarial view of loyalty — either the client's decision is fully accepted or the engineer is being disloyal. It ignores the professional's position of holding several differing responsibilities — principal amongst them being his *duty* to the public, his personal integrity, and his *obligations* to the client.

Loyalty implies a sovereign, the client is not the engineer's sovereign. The engineer is an independent agent, not an obedient servant. Such freedom is one of the key benefits of an independent professional. He can often view things more objectively than those who are under stressful management conditions and thus may be able to safeguard the public and thus all parties concerned against major disasters. The professional independence of the engineer is ultimately to the good of the organisation and should be encouraged among engineering employees, and required of consultants.

Laws are frequently drafted only after the occurrence of serious incidents, and are often clumsy in operation. The public more often relies upon the independent and aware professional to ensure that sound and balanced decisions are reached. Independent reporting is relatively rare. If such action is responsibly handled it can even have the effect of reinforcing trust. If it can be seen that there are means of ensuring against partisan decision making, society will be reassured. If both management and the professionals recognise this fact, they will be less inclined to ignore professional advice.

A checklist of actions which the engineer should consider if he thinks there is a problem, and that external reporting is called for, might include the following.

- Confirming that the risk to the public or to fellow workers warrants some action. This requires a careful sifting of the evidence to assess if the harm to the public is significant.
- Motives, personal and corporate, should be carefully examined at an early stage. A colleague may be able to help in the assessment of the decision making process. The dangers to be avoided are either the heroic motive, with the engineer seeing himself as fighting a battle, or the revenge motive.
- The evidence should be verified and well documented, with records kept. Once more this may mean consulting with other colleagues, in gathering or analysing data.
- Determining the area of concern and identifying the appropriate referral agency. Establishing which agencies may have the authority to enforce rules and laws related to health and safety, race relations, etc. At this point the professional body might be asked to help to clarify the issues.
- Stating objections clearly. Avoiding confrontation if at all possible. Being assertive rather than aggressive. Sticking to the facts and reasserting them if need be.
- All standard procedures should be followed. Supervisors and management should be kept informed at all times.
- Before standard procedures are bypassed, the issue should be presented to the person most concerned with the matter. If it is clear that the evidential material is on record and has been verified and confirmed by independent sources, it may be possible to obtain a positive response. This may involve working through the decision making process with those concerned.
- Anticipating and documenting any retaliation. The likelihood of some retaliation may be high.

Such a check may reduce the need to report the event beyond the immediate organisations concerned, and represents the professional preparation necessary before reporting to an extra-organisational authority. The checklist reinforces the need to avoid conflict throughout the process.

If any dispute between the engineer and employing authorities escalates to public disclosure and the involvement of the press, all parties are liable to lose.

More rarely, situations may arise where the professional engineer becomes aware of ethical issues only after decisions have been made and implemented. If these involve unsafe or unacceptable practices then the appropriate public authority should be advised, and the responsibility passed over, going to the highest level of authority if need be. Similarly, if corruption or illegalities have been disclosed then these also should be reported to the appropriate responsible legal or public sector authority.

It is in the interest of all concerned to have established protocols that enable those involved to address the issues, some of which may start off as expressions of concern, rather than of complaint. Such a protocol requires carefully thought-out complaints and consultation procedures. Such procedures can create, for out-of-line reporting, accepted structures in their own right. They would include the following.

- The procedures should stipulate the steps to be taken after the reporter has been unable to take the matter further with his immediate senior manager.
- Recognition that matters are not limited to the individual's own area of responsibility. This provides a reminder that the responsibility of a professional to the safety and well being of society go beyond particular role responsibilities.
- An initial review of the data by competent professionals in the organisation. This may be supplemented by the appointment of an independent reviewer, or if serious, of a panel of reviewers, to determine if there is cause for concern and if any more data is required. Responses should be given speedily and communicated to the reporter with the reasons for any decisions reached.
- Written records of all transactions should be kept for internal audit, or where a government body might be involved for public inspection, the record should document the concern, the initial response and the assessment of the consequences if this concern is not addressed.
- To assure any originators that the matter is being taken seriously there needs to be a right of appeal to another authority if the finding

does not meet the concerns of the reporter. At this point the professional body might be involved, perhaps with a member from a selected panel of senior engineers, who could be given evidence in confidence. The results of this more formal approach would be communicated to the reporter within a defined period.

- There should be the greatest possible transparency at all stages, including prompt feedback to the complainant.
- This whole protocol should be clearly set out in the staff handbook.
- The whole process should be reviewed periodically so that the effects of good practice can be seen and any bad practice modified.

The new ICE Code encourages the development of such procedures (see clauses 1, 2, 3, 5, 7, 10, 11 and 16).

4.3. Sustainability — engineers and the environment

Over recent years there has been a great increase in the understanding of sustainable development and 'green' issues, but little has been published on the changes needed in the construction industry to accommodate these issues. Engineers have, by tradition, been at the heart of efforts to increase levels of production and consumption, generally as a means of improving conditions and servicing the growth of economies.

The new mood is one of conservation and sustainability. The construction industry, together with government and responsible clients, is now beginning to address these issues. The changes of attitude needed to implement these issues in the processes of design and construction are taking place These changes are especially important in civil engineering where the works have the potential to affect not just the current generation, but many future generations, in terms of aesthetics, quality of life, environmental impact and investment in the future. In the development of major projects, engineers are called upon to make decisions which can have far-reaching effects, not only for those immediately concerned with the project and its aims, but for future generations. In the case of some major projects, their effects have an international if not global significance, affecting the environment for many years, if not centuries.

Such decisions are of direct concern to the engineer responsible for the design and execution of projects. In order to achieve an end which seems to offer some general 'good', we frequently have to incur some more limited 'bad'. This is the ethical problem of 'the conflict between rival goods'. The frame of reference given in Chapter 2 — (Figs 2.5, 2.6) can assist in evaluating the relative merits and demerits of alternative decisions. Clearly, some decisions are very limited in their consequences, and the big questions are not relevant thereto. Our personal preferences may have little effect internationally.

Engineering responsibilities for the environment are currently governed by laws controlling environmental matters. Stringent regulations now exist at local, national and international levels. These are designed to ensure that the industry conducts its business in an environmentally friendly fashion. British Standard 7750 — *Specification for environmental management*, ISO 14001 — *Environmental management systems: specification with guidance for use, and Environmental assessment directive* 85/337/EC all provide guidance and control to ensure that projects receive an appropriate level of environmental assessment and analysis. More importantly, these regulations have begun to form a culture in the industry where environmental awareness is at the heart of the project, rather than a late 'add-on'.

The far-reaching effects of some actions — deforestation for example — require the highest level of consideration. This consideration of the wide-spreading consequences of some major engineering works is reflected in the increasing call for full environmental impact assessments for all major projects. The EEC has led the way in this call. The main headings under which such assessments are to be made are given in Table 3.3, Chapter 3. These are incorporated in the ethical audit tables in Chapter 6, and illustrated in the case study set down in Appendix 1.

Engineers find themselves attempting to rationalise/comply with many conflicting needs and requirements when attempting to realise business projects. Clients may have many outlooks with regards to the environment.

- They may be part of an organisation which intends to 'pay the fine' rather than attempt to use resources on environmental protection considerations.
- It may alternatively be part of an organisation that adopts governmental restrictions as a cost of doing business — but complies without enthusiasm or commitment.
- They may be part of an enlightened and responsible organisation with a wish to incorporate the best environmental considerations.

In these situations the engineer is confronted with his own duties and responsibilities. 'A member shall have full regard for the public interest.' (3)

The engineer is therefore committed to a concern for environmental protection and even improvement. Little direction is currently given as to how this concern should be addressed or implemented.

> If a clean environment is promoted to protect human health, how do we measure 'clean'?
> If a project includes the construction of a dam which will destroy a 'natural' river and flood farmland, how should we regard that destruction of a natural environment or even the viability of the overall project in terms of good or bad social policy?
> Should the engineer act and object as either his professional or social ethics demand?
> In what capacity should he object — professional or individual? What happens if the ethics of these differing value systems are in conflict?

The ethical review of projects involves complex, sometimes subjective qualitative analyses, cost-benefit analyses may be almost impossible to quantify. The analyses may produce conclusions which, while justified in engineering terms, offend the moral beliefs of 'ordinary people'. The conflicts between personal and professional systems are covered in Section 4.2 of this Chapter, covering professional responsibilities for reporting actions believed to be detrimental to society or the environment.

The actions of the engineer on matters of sustainability and environmental impact are summarised below.

- Engineers must assume a responsibility for the effects of their work and endeavour to make a substantial contribution to the protection of the environment.
- They must not participate in projects that are unnecessarily destructive to the environment — even if these projects do not endanger physical life or health.
- Engineers should express their professional opinions on environmental matters, based upon sound knowledge and analysis. They must take care that personal protests do not overtly conflict with their clients' wishes, since this would invite public disrespect and mistrust for the profession, violating Rule 2.
- Where clear guidelines are available, such as in the EEC Directive 85/337/EC (Table 3.3), the engineer should observe these and not seek to impose more stringent conditions. Where no national guidelines exist then the engineer has the responsibility to set

appropriate standards, to communicate these to others, and to work in accordance with these self-generated regulations.

- Should the engineer consider that the project is unnecessarily harmful to the environment then he should attempt to explain the situation to his clients. In the event of his opinion being disregarded the engineer has the right to withdraw from the project, with the right to make his concerns known to the relevant authorities.

4.4. Bribery and corruption

4.4.1. Definitions

Bribery — to promise or give something, often illegally, to a person to procure services or gain influence.[1]

Corruption — making or being corrupt, dishonesty especially bribery.[2] Misuse of public power for private profit.[3]

4.4.2. International policy

Engineering projects are complex and frequently high-value enterprises. They involve the participation and co-operation of large numbers of people. They also present many opportunities for bribery and corruption.

La Fédération Internationale des Ingénieurs Conseils (FIDIC) has noted the apparent increase in corrupt practices and in its Policy Statement on Corruption (June 1996) advocated clear guidelines. FIDIC defines corruption as 'the misuse of public power for private profit', arguing that it is morally and economically damaging.

> Corrupt practices can occur at all stages of the procurement process: in the marketing of engineering services; during the design; in preparing the tender documents (including the specification); in pre-qualifying tenders; in evaluating tenders; in supervising the performance of those carrying out the construction; issuing of payment certificates to contractors; and making decisions on contractors' claims.
>
> Corruption, definable as 'the misuse of public power for private profit' is morally and economically damaging. Firstly, it jeopardises

---

[1]   *Collins English Dictionary.*
[2]   *Collins English Dictionary.*
[3]   FIDIC.

the procurement process, is always unfair and often criminal. It saps money from required development projects and adversely affects quality. Secondly, and worse than being pragmatically wrong in allowing wasteful procurement, corruption is more basically wrong because it undermines values of society, breeds cynicism, and demeans the individuals involved. It is more than stealing funds, it is stealing trust.[1]

While the FIDIC statements of policy are clear and unambiguous, in practice moral problems in encountering and dealing with bribery and corruption often seem to be complicated by the context and magnitude of the gifts or bribes involved. It may also be the case that these practices are not unusual or even illegal in some cultures.

For instance, an engineering practice, having established a successful record of working overseas, may be faced with problems when attempting to seek work in new areas of the world. An engineer may be advised by high-ranking governmental officials that it is 'normal practice' to make personal gifts to those governmental officials who have the responsibility to award contracts — without these gifts there will be no opportunity even to be asked to tender for the work. Discreet enquiries often suggest that competitors have accepted local practices of this kind. The engineer is immediately faced with a dilemma that will, if he behaves in an ethical manner, effectively prevent him from working in that country.

Alternatively, the engineer may be faced with a similar, if smaller, request for payment from a lower-ranking official. He may then reflect, in this case, on the norm in his own country which would permit him to exchange gifts or hospitality or entertainment of a similar value with his clients or suppliers, without the fear of encountering the same dilemma. He may be confused. How are these cases different? Should he consider falling back on the 'When in Rome . . .' rationale?

His own institutional code may guide him in the situation but it is more likely that he will require guidance from a set of additional criteria. No single rule will cover all eventualities in these cases. He must decide for himself whether the gift is an inducement intended to influence an independent professional judgement or simply an expression of friendship or social custom. Obviously the value of the gift is relevant but this must also be related to the circumstances and local environment in which the gift is offered or accepted.

It is often impossible to read the mind of the donor or receiver in these cases, though morally the engineer must try to do so. Once he has done

---

[1] FIDIC. *Policy statement on corruption*, June 1996.

this he must make his interpretation of the situation clear to the other party in the relationship. It may be that he must apply a test based on what 'reasonable men' might infer from public disclosure of the nature or size and circumstances of the gift. Does the giving or receipt under these public conditions infer a position of obligation, ulterior motive or raise the suspicions of 'reasonable men'?

Under these circumstances the implications of a 'When in Rome' policy become clearer. The code under which he works must be read not only literally but also in the spirit of its purpose — to uphold the integrity and standards of the profession. Anything else is a rationalisation of the true conditions.

4.4.3. FIDIC standards

The high ethical stance of FIDIC sees the engineer having a responsibility for maintaining moral meaning and a moral framework in society which goes beyond the confines of work and the professions. As a result their recommendations go beyond the responsibilities of the individual engineer and provide overarching guidelines for professional bodies and companies.

The recommendations include the following.

- In implementing particular projects, consulting engineers should recommend to their clients the most appropriate and objective procurement process or delivery system, consistent with the demands of the project.
- Funding agencies should be kept fully informed by the consulting engineer of the procurement steps as they occur. The consulting engineer should notify funding agencies of any irregularities, in order that cancellation or other remedies may be exercised, in accordance with the loan agreement.
- Consulting engineers should be aware of local law regarding corruption and should promptly report criminal behaviour to the proper law enforcement authorities.
- FIDIC member associations should take prompt disciplinary actions against any firms found to have violated the FIDIC Code of Ethics. This could include, among other actions, expulsion and notification of public agencies. Procedures should be established by member associations to ensure that due process of law is afforded in such cases. The procedure to determine whether the expulsion of a member firm is warranted, should be conducted confidentially but expeditiously.
- Member associations and their members (firms and individuals)

should internally develop and maintain systems to protect their high ethical standards and codes of conduct. They should cooperate candidly with other organisations which seek to reduce corruption. Member firms should associate themselves only with other member firms who share similarly high ethical standards.

- Member associations should foster and support the enactment of legislation in their own countries, aimed at curbing and penalising corrupt practices.

Such recommendations reinforce the view that in addition to the development of ethical decision making the practice of ethics demands that groups work together to develop standards, and that ethics be seen as a function of partnership not simply the judgement of an individual.

4.4.4. Ethical decisions

In reality the overall conclusion seems to be that the engineer will be called upon to make his own assessment of the situation in the majority of cases. It is important however that he be objective in deciding what is happening and the real intentions behind the event. Only with this overview can he make his decision, guided by his own ethical code and that of his institution. On this basis he can negotiate the 'rules' under which he will operate with the other parties in the relationship.

Corruption and the associated attempts to stamp it out are nothing new. However, there remains the problem of how to deal with those cases which fall within the 'grey areas', these often being where the engineer or manager work. A good example of this is the situation which arises when dealing with agencies from other countries who may demand undisclosed payments.

There are three traditional approaches to this.

- *Ethical conventionalism*, which argues that the conventions of the country be followed — 'When in Rome, do as the Romans do'.
- *Ethical fundamentalism*, which demands that the same high ethical standards be applied across different countries. This is the logical outcome of the FIDIC recommendations.
- *Ethical case work*, which argues that each case be carefully examined.

The approach of casuistry would seem to be the most pragmatic and effective one to take for the individual manager or engineer, not least in an area where there is so much disagreement even amongst ethicists. However, even this should not be seen as the procedure simply for an individual to follow. The test of reasonableness is too prone to vary with individuals.

This practice of 'casuistry' involves several stages.

- *Clarification* of any act with the donor or donee. This applies to the local as well as international situations. Lack of clarification can lead to unfounded assumptions, not least on the part of givers of gifts.
- *Accountability*, such that all aspects of gifts/payments are known by the firm and can therefore be called to account.
- *Co-operation* with all agencies who have a concern for the area of business, from government to other industries, to international agencies. This ensures the build-up of consistency in practice.
- *Examination of the context*. Is the act actually one of bribery? Does it involve a breakdown of trust or is it actually a form of tax which is accepted by all involved or which supplements the low pay of officials? There are some rules of thumb which might pass the test of the 'reasonable man' and help to define what might be bribery, including the size of any 'gift' or the past record of the firm in question and the context of the transaction.
- *Examination of the options* and the possible consequences. Could standing out against the practice actually lead to a general improvement in practice both in business and in the country involved? Could it lead to bad consequences for the company and possibly for other companies involved?

To these and other questions there may be very different answers for each situation.

All of these approaches are designed to increase transparency and make it less easy for abuses to occur at any level. In the final analysis the engineer might also have to examine his own attitudes toward the case and ask how far his judgement might be affected by any approach. Where there is doubt the transaction should be avoided.

4.5. Public policy — design, build, finance, operate projects

The 1980s and 90s have seen great changes in the construction industry, and there are no doubt other changes to come. A new type of client has emerged, together with new types of consultant and contracting organisations. This is, in its turn, changing the nature of competition, procurement and technology.

The power generators, British Rail, the water authorities, and other major organisations all now operate in the private sector. They are using their new power and size to demand more competition, more efficient and effective procurement, and radical methods of financing for construction and operation. These customers, together with some major public sector organisations — hospitals, universities, highway authorities, etc. — are transferring more responsibility to contractors and consultants by using design and build construction contracts and fitness for purpose specifications. There has been a general move for consultants away from the traditional status afforded to a professional representative, an agent, to a much more contractual relationship, with an attempt to create more sharply defined scopes of work. These are being seen as ways of overcoming the 'construction claims culture' which was felt to be exploiting the traditional interface between design and construction. Construction projects have become larger, many in excess of £100 million.

The UK government's Private Finance Initiative (PFI) was seen to offer the potential for broader and improved public services as a result of partnership of the private and public sectors.

In a letter to the Chairman of the Civil Engineering EDC, in October 1986, the Prime Minister (Margaret Thatcher) wrote:

> . . . *The Government welcomes the use of private sector finance and expertise in improving the enterprise and management efficiency with which services can be delivered.*
>
> *Privatisation, that is transferring the responsibility for providing service . . . wholly to the private sector, is the most complete way to secure this. Where the public sector retains responsibility to provide a service, other ways of getting the private sector to provide an input to that service, such as contracting out, can be valuable where they are more cost-effective than provision from within the public sector.*
>
> *Similarly, where the public sector would otherwise have invested in a capital project as part of the provision of a service, the introduction of private finance for the capital project is welcomed, if the proposal is more cost-effective.*

The PFI of 1992 is now being used for many major projects. These privately financed projects require a new type of procurement consortium capable of providing large sources of finance for the development, management and successful completion of the projects. The increasing scope of projects has resulted in more flexible contracting organisations, with consortia being created for specific projects, with a finite life.

The changes in the philosophy of the promoters of these contracts has affected the manner in which the players in the contracts interact and behave. Traditionally the promoter would seek the lowest cost solution to his problem, often separating considerations of maintenance and operation from the initial construction costs. This separation sometimes resulted in long-term disadvantages in the life-cycle costs of the project. Contractors in these conditions did not seriously attempt to reduce operational or maintenance costs. Turnkey contracts, or design/build/finance/operate contracts (DBFO), attempted to overcome this attitude; the promoter — now the contractor — having to consider the design, construction and operation of the project. Ways of improving life-cycle productivity become important. 'Partnering', collaboration between public and private sector organisations, is actively being promoted by the industry, enabling public bodies to 'outsource' their specialist services, rather than keeping their own in-house professional teams.

As a result of this initiative the private sector has become involved in a broad range of projects in the areas of information technology (IT), health care, transport and highway provision, defence, prisons, educational building, operational and maintenance projects and airports. It has also become involved in the recently privatised industries from which it was previously excluded when these industries were state owned and operated.

Whilst this private sector involvement has undoubtedly resulted in the more efficient provision of services, many misgivings have been raised, some even querying who are the ultimate beneficiaries from the PFI system. Deregulation and privatisation, both in the UK and worldwide, have tended to swing the balance of power away from governments. An elite and relatively small group of major transnational companies have been shown to have the financial resources and technology to take on operations of this scale. Many of these companies have similar investments throughout the world and the driving force is perceived to be successful investment — sometimes at the expense of users whose points of view or interests may have been considered by public sector managers on the basis of service, rather than market considerations. The World Bank has commented on a 'worsening of the already skewed concentration of ownership distribution'.

The general underlying problem can be seen in a speech made by a representative of a UK power supplier.

> *Firstly we focus on operating profit, seeking ways to build cash inflows by maximising revenues and reducing costs. Secondly we focus on capital employed to optimise cash outflows, looking very carefully at the timing and extent of investment and taking a rigorous view of the disposals. Finally we focus on understanding the cost of capital and the implementation of balanced financing policies . . . This focus is in marked contrast to the priorities that the company had prior to privatisation (when) our*

primary role was the maintenance and security of supply to our customer base.

Our approach to investment was to invest whatever was considered necessary on technical grounds to deliver an electricity supply to our customers.

Statements like these reinforce a public belief that private sector business behaves in a more questionable manner than the public sector because it appears to be less accountable to the people. Lord Nolan's Commons Commission has not shown this to be true, but there remains a fear that privatised industry will cut sections of business that are not profitable. In terms of environmental considerations privatised industry may well want to increase profits and consumption rather than conserve resources.

The problem of ethics is not, therefore, one of fraud or bad dealing but a clear issue of whether or not the market alone should determine what services are offered, which should flourish and which fail. The duty to consider the stewardship of the company and the common good of the minority is eroded.

In such projects the ethical issues are concerned with the problems of providing and controlling the general public services essential to the smooth working of a healthy community and economy. The efficiency of a management not locked into a state bureaucracy is undoubted, free to make decisions on a day-to-day basis and changing policies to suit changes in demand or structure. The aims of such an organisation may not, however, be primarily to deliver a public service. The state has attempted to control such problems arising from the return of some facilities — considered for many years to be a state responsibility — to the private sector, by establishing national regulating authorities, charged with monitoring the management of the facilities. Such organisations came under the terms of reference of the 'Nolan' Committee as 'Quangos' — Quasi Non-Governmental Organisations — including such organisations as hospitals, universities, etc. (see Table 3.2).

A wholly satisfactory 'culture' for such organisations has yet to evolve, covering as it does everything from the National Lottery to the nation's power supply. The professional engineer may be concerned both with serving a PFI or operating unit as a member of a private organisation, or, occasionally, as advising one of the Quangos. In either case the independence and integrity of the profession are of great importance. Clauses 1, 2, 3, 4, 5, 7, 11, 13, 14 and 16 of the ICE Rules of Conduct are all relevant. In the absence of clearly defined legal or traditional frameworks of action the responsibility of the individual to be self-regulating in his standards is of particular importance.

4.6. International issues

4.6.1. Globalisation

We live in an age when technical and political changes have been taking place more rapidly than at any time in the past. The scale and range of societies, the size of projects undertaken by them — projects concerned with education, health care, government, public works and many other areas of communal concern — have greatly increased over those undertaken in previous centuries. Perhaps the rate of change is even more significant than the scale of change. In addition to the scale and rate of change, and the changes in population distribution, we have to recognise the nature of the change. Technological developments have placed vastly increased sources of power under the direction of mankind, and have very much enhanced the speed with which we can communicate with one another. Within decades the time taken to communicate with some parts of the world has been reduced from months to seconds.

In parallel with this development there has been an unprecedented increase in the population of the world, requiring quite different social relationships and dependencies than has been the case in the past. Each of these factors affect our traditional values. The behaviour of men and women in authority is a reflection of the values and behaviour of the societies in which they operate.

Until the second half of the 20th century most organisations operated within circles that shared their cultural and social structure — often a single cultural framework. Operations outside this sphere were seldom directly influenced by the different cultures or the values encountered. Indeed other value systems were not generally considered or even observed. As a result companies and nations became or remained 'closed'.

In the recent past, technology and media expansion have 'shrunk' the world and society, reinforcing the concept of all people having common concerns, and needing to reconcile differences in their value systems. These changes in outlook have been reflected in significant international high-level discussions driven by the recognition of common concerns of all national communities. In recent years there have been global summits on topics such as: the environment, 1992; human rights, 1993; population, 1994; social development, 1995; women, 1996. These summits have provided a mechanism by which the aggravating effects of shrinking resources, sustainability and expanding demand can also be addressed.

The globalisation of markets has been accompanied by the development and growth of multinational corporations (MNC) which operate across local and national boundaries. These businesses have become international, or indeed supranational, more rapidly than many of the national communities in which they operate and they have influenced the nature of the development cycle by their presence. The increasing international flows of capital, technology, trade and people have had the effect of changing the nature of local organisations, governments and people of countries, and have led to social changes and development.

The importance of the standards and values of the decision making parties becomes increasingly significant. Such executives may have to be almost entirely self-regulating. It is therefore even more important than has traditionally been the case that we understand the reasons behind our regulations. The identification of a group of executives, involving the most senior management of the project parties, to consider these matters can be very important (see the case study on the Lesotho project in Chapter 5). Such a group may benefit from the inclusion of distinguished, independent, lay members experienced in the structure of the societies involved and in their traditional values, but with a global view of the needs and the nature of mankind.

The influences and effects of global enterprise and competition on exporting and host countries can be far-reaching. It is important that respect for cross-cultural differences and concern for human as well as business interests is developed and maintained. It cannot always be assumed that *we* know what *they* want and what is good for *them*. Mutually acceptable aims, values and standards are vital if economic success and public acceptance are to be assured.

4.6.2. Cross-cultural issues

As has been stated, world communities are much more interdependent. Differing views as to the nature of the 'good' and of the 'right' have a real impact on peace and prosperity. The decisions of a single powerful individual can affect the lives of many millions of others — for better or for worse.

The issues arising from this globalisation are complex. They require an in-depth study of the value systems and the physical, economic, political and cultural parameters of all the individuals and communities involved. This requires a cross-cultural ethical audit of a much more strategic nature than the audit of a particular project or enterprise.

> Saudi Minister
> *It is not good practice to add your relative to the tender list.*
> *Why not? I can trust my relative to perform. He is honour bound*
> *so to do. Why should I trust your choice of an unknown foreign*
> *contractor more?*
> The relative formed a consortium with a competent international firm — but its bid was unsuccessful.

The professional engineer may have neither the training nor the experience to carry out such a study. It may, however, be within his authority to ensure that all these factors are considered, and to engage the assistance of fellow professionals who can make an objective and understanding assessment of the situation.

The operations under consideration may involve a high degree of risk. They may be subject to fluctuations in the money markets, or instability and disturbance in some of the nations involved. Local limitations may make it difficult to carry out some of the tasks. There may be no local legislation to control contractual relationships or health and safety aspects of the work, as well as to monitor environmental impact.

> Gulf Arab Client
> *Your choice of (an Arab) supplier will cost you more*
> *How else do you think I distribute my wealth?*

An example of the relevance and importance of understanding the differing standards and development of differing countries can be seen in the problems of transferring technologies. As a guide to assessing the needs and likely consequences of such transfers it may be helpful to carry out a study of the context of any proposed transfer. The factors to be appraised, and suggested considerations to ensure successful integration of the two or more cultures, are set out in Table 4.1. These are divided into two categories — cultural and non-cultural, which are again divided into two types — human and technological. Factors are suggested that will assist in reconciling the differences. Staff preparation by training in the recognition of important factors, and in adopting practices that smooth the transfer of technologies, is essential.

With this basic research and sympathetic understanding, the integration of technologies and methodologies becomes much more effective and acceptable. The essence of the transaction is to serve, not to make others

Table 4.1

Cultural factors	Non-cultural factors	Transfer factors
Human		
Education	Talents	Personnel selection
Experience	Motives	Special training
Tradition	Needs	Wide experience
		Open motives
Law		
Convention		
Geography (sociological)		
Technological		
Training	Knowledge	Research
Skills	Science	Analyses
Organisation/methods	Basic principles	Appropriate technology
Materials		
Geography (physical)		

like us. One of the most rewarding aspects of the engineering professions is the enabling of activities that improve the quality of the lives of others, without diminishing the rich variety of human experience and fulfilment.

Figure 4.1 illustrates in diagrammatic terms the process of integrating a new technology with an existing culture. The need to identify the factors that may become conflicts of interest within traditional value systems is an ethical responsibility for the professional. Undertaken conscientiously and intelligently, such an exercise will, by anticipating and preventing conflicts, prove valuable in reducing delays, in accommodating changes of design and construction practice, and in ensuring collaborative working between people of differing cultural and experiential backgrounds. But being transnational, it is likely to lie beyond the reach of national laws and professional codes — individual self-regulation is again of prime importance.

4.7. Conclusion
Responding to ethical dilemmas in the engineering context cannot always be straightforward, not least because life itself is rarely so. Hence, handling such dilemmas requires

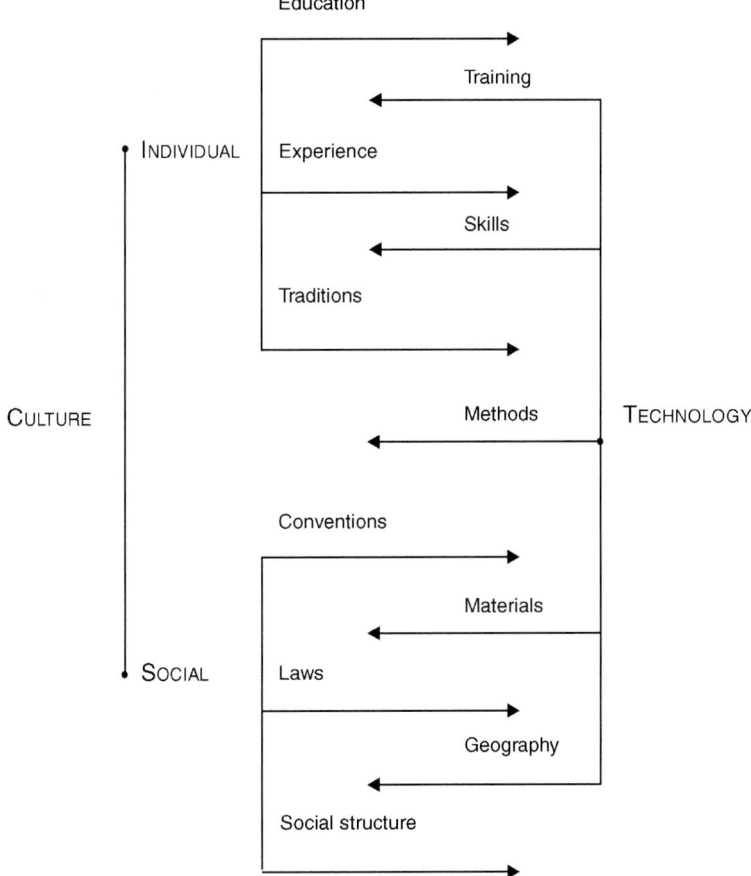

Figure 4.1. Transfer of technology

- ethical thinking which enables the professional to understand the data, values and consequences of the case
- careful attention to methods, so that the dilemma is not turned into a crusade — social and self-awareness and the development of a professional value system
- awareness and development of codes of ethics in particular companies, or on a particular project, so that ethical discussions are open and evident
- awareness and development of interdisciplinary codes and cross-cultural factors.

Because no simple rules will deal with all ethical problems the exercise of responsibility becomes central. This involves establishing the responsibilities to the several parties concerned, which requires negotiation, to ensure that the professional does not take on too much responsibility. The professional must, however, take responsibility for ethical dilemmas and work through the ethical decision making process — avoiding any denial of responsibility. Such responsibilities cannot be confined to particular roles. The responsibilities of the engineer and the engineering profession include those to society in general.

5

Case studies

Duty is the action proper to each man which keeps to what is fitting and honourable as circumstance, person, place and time require.
Marsilio Ficino, Renaissance philosopher – c.1450

5.1.   Introduction

The case studies given in this Chapter were selected not because they were of outstanding interest, but in demonstration of the kinds of ethical decisions that occur commonly on civil engineering projects. The descriptions are not intended to be comprehensive, but simply to focus attention on the issues or procedures being considered. They represent a wide range of topics.

The notes on the Lesotho and Channel Tunnel projects demonstrate how valuable it is to have a formal, established, system of consultation and communication as a means of identifying and avoiding possible conflicts of interest on large complex projects extending over a considerable period of time. In the case of the Channel Tunnel, the long evolved and established system of approval by Parliament, representing the interests of the community, with its system of public select committees, served this purpose. In the case of the Lesotho project, a consultative system was created specifically for the project, since no formal system was in place. In both cases time was set aside purely for the consideration of social and environmental issues.

5.2. Citicorp Tower, New York

William LeMessurier is an eminent American structural engineer who has developed an enviable reputation and experience of the design of skyscrapers, following his work on the Boston State Street Bank and the Boston Federal Reserve Bank. He was retained by Citibank to design their corporate New York headquarters, the Citicorp Tower. The building was completed in 1977. His design was radical, incorporating nine-storey high stilts located in the centre of each building elevation — rather than at the building corners — so that a church could be located beneath the new bank structure. The structure was based upon LeMessurier's diagonal bracing design. This building form resulted in great savings in weight over contemporary structures and an innovative system of tuned mass damping — a 400 ton concrete block located at the top of the building — was incorporated to help the building resist sway caused by wind.

By 1978, LeMessurier was engaged in the design of a similar building in Pennsylvania. During the course of this work he discovered that the contractor for the Citicorp Tower had made changes to the initial design, during the construction phase. Bolted connections had been substituted for the full-penetration welds specified for the bracings, during the course of construction of the tower. At that time LeMessurier did not consider that the changes had any effect on the structural stability or safety of the tower.

In June 1978, as a result of queries from academic sources, he made a check on the design of the tower. Where the original design had been made on the basis of wind forces applied at right angles to the face of the elevations in accordance with contemporary New York standards, his new check considered wind quartering the building on a diagonal.

He found that with quartering wind loads, the stress in a number of members increased by 40%. A 40% increase in these members could result in a 160% increase in the stresses at their joints. The check revealed that too few bolts had been used in the connections.

Original wind-tunnel tests were updated and further wind-tunnel tests carried out. LeMessurier's worst fears were confirmed; the tower was at risk from total structural collapse under the loading from a 16 year storm. A structural solution to the problem was not difficult. The joints must be made more robust, but because the building was occupied, the modifications would be difficult to undertake. The modifications would also have to be carried out before the start of the rapidly approaching hurricane season.

LeMessurier knew that disclosure of the problem would undoubtedly damage his public and professional reputation but he contacted representatives of his client and a meeting was arranged for 31 July 1978. Citicorp was informed of the problem; the dangers of collapse, the method of strengthening the building and the likely cost, were discussed. Shortly afterwards a plan for the emergency evacuation of the building in the event of a hurricane was discussed with city officials. A press release was issued stating that the building was to be refitted to withstand higher wind loadings. An unexpected reprieve from embarrassing press attentions was given by a local press strike.

Repairs began with joints being exposed and protected behind enclosures whilst work was carried out. Welding was carried out after office hours when the building was empty. A scare when a hurricane was predicted to reach the building came to nothing because it turned away before reaching the city. The work was completed by the end of September 1978.

LeMessurier obviously feared for his career, but at no time did he allow his apprehension to interfere with his work on the tower-strengthening programme. On completion, Citicorp requested that LeMessurier reimburse it for the repair work (estimated to have cost between $4 and $8 million). His insurance company offered to pay $2 million and Citicorp agreed to accept this amount, to find no fault with LeMessurier, and to close the whole matter.

It was anticipated that LeMessurier's insurance company would radically increase his liability insurance premium for future work. However, it was agreed that he had acted in an exemplary manner and that having solved the problem, a major disaster had been averted. As a result of his behaviour throughout the affair, LeMessurier emerged with his integrity and reputation intact.

It is reported that, rather than increase his premiums, his liability insurers actually reduced them.

5.3. Miscellaneous small cases

5.3.1. Transfer of commission

A consultant (B) was asked to undertake work on a listed building. Prior to his appointment, some work had been done by another consultant (A), who had not been paid. The client asked consultant (A) to sell his work to the new appointee, (B). The asking price was high. (A) was said to be concerned about his residual responsibility. The client then asked (B) to negotiate a purchase for (A) and to take over the responsibility for any latent defects. Are there any ethical problems?

Since all parties are aware of the situation, and (B) did not deliberately attempt to usurp (A) there is no breach of the Rules of Practice. In the circumstances, however, it would be advisable for (B) to ensure that his insurers are willing to have him take over the risk, and that his agreement with the client ensures that he be paid his fees, even if the project does not go ahead.

5.3.2. Conflicts of clients

A consultant was appointed by a client to investigate structural damage caused by subsidence of his property. The client's insurers appointed their own loss adjusters as their consultants to carry out further investigations and undertake designs for remedial work. It has been the case that sometimes the loss adjusters have offered financial benefit to the insured by organising payment directly from the insurance company, with the excess to be paid in the contractor's final account, rather than upfront. In such cases the original consultant must be paid off by his client at the outset of the remedial work. The owner should be able to claim his consultant's fees from the insurers. If the loss adjusters' engineers are members of a professional institution then they may be in breach of Rules 2, 5, 14, 15 and 16, but only if they have deliberately sought to replace the original consultant. It is standard practice for insurers to have a panel of approved consultants from which an appointment is made to carry out the work. The use of such a panel is not unethical. In handing over any work to a second consultant, care should be taken to define the transition clearly, and to ensure that the replacement consultant accepts responsibility for any work undertaken by the original consultant.

5.3.3. Definition of responsibilities

A local authority is engaged as an agent to act on behalf of the Department of Transport. The authority is to carry out the design and the supervision. After due process a contractor is appointed to carry out the work.

The authority itself subsequently undertakes a subcontract to carry out some testing work for that contractor, for a fee. Is there a conflict of interests?

A conflict of responsibilities is clearly a possibility. Should there be any fault on the part of the testing organisation, the responsibility is clearly divided between the client's agent, the authority, and the contractor's subcontractor (also the authority). The situation should be avoided by requiring the contractor to obtain his testing services elsewhere (see Rules 1, 3, 5, 11 and 14).

5.3.4. Ownership of bid designs

A client engages a consultant who prepares a design for a project and issues the design for contractors to price. One contractor, with or without the assistance of a consulting engineer, identifies a better concept for doing the whole or part of the project than that proposed in the original design. This is then submitted along with the contractor's bid as an alternative proposal.

The client likes the alternative proposal. However, he does not wish to enter into a negotiated contract with the bidder, because he may be able to obtain a better price for the alternative, from other contractors.

The alternative scheme is not necessarily covered by copyright because it is a concept only and in any case, minor changes would probably avoid any breach of that copyright. The consultant is therefore asked by the client to prepare an alternative scheme and obtain fresh tenders.

What is the correct ethical stance? The situation can be avoided by express conditions in the tender documents, either prohibiting alternative designs, or setting out clearly the method that will be adopted in dealing with them. Alternative schemes prepared by contractors are the property of the contractor. To use such a scheme requires an acknowledgement of the contractor's ownership, with due compensation being paid if it is used. If this is not the case then a negotiated contract is the correct course of action, holding the other tenders until satisfied that the negotiation is really to the advantage of the client. If the contractor is not to be used then he must be compensated for the preparation of his scheme. It would not be ethical for the consultant to use the scheme knowing that the client has not obtained the contractor's permission for him to do so (see Rules 1, 10, 11, 14 and 15).

5.3.5. Responsibility for public safety and interests

A consultant was called in to advise a statutory authority on the factors responsible for the failure of some vibro-placed stone column foundations located in a backfilled sandpit. It was clear on subsequent inspection that there was a risk that some landfill gas was present. This would not have been evident at the time of construction, given the knowledge of gas generation at that time. This risk might affect some properties which were not the subject of the present commission, and were not included in the consultant's brief.

It is clearly in the public interest that this issue be discussed and resolved, and the consultant was duty bound (Rule 3) to report the matter to his client, even if this meant that his client might be faced with considerable extra cost in resolving this unforeseen hazard. The client was informed, tests were carried out, and the problem turned out to be not as severe as it might have been. The engineer's responsibility cannot be limited by his contract if he becomes professionally aware of other significant issues (see Rules 1, 2, 3 and 11).

5.4. The Lesotho Highlands Water Project
This project provides an interesting example of the methods by which the environmental and social aspects of a major project can and should be included in the project management arrangements of a major programme of construction works.

5.4.1. Project background.
The Lesotho Highlands Region covers one third of the area of the Kingdom of Lesotho. This small country is completely surrounded by the Republic of South Africa (RSA). The Region is the source of the Senqu/Orange river, the largest and longest river in the area.

The Kingdom of Lesotho is poor and relatively undeveloped. Many parts of the highland region are inaccessible except on foot or horseback.

The primary objectives of the Lesotho Highlands region project are

- to redirect some of the water which presently flows out of Lesotho northwards towards the population centres of the RSA
- to generate hydroelectric power in Lesotho using the redirected flows
- to provide regional social and economic development, water supply and irrigation in Lesotho.

These objectives are to be achieved by means of the construction of a series of dams, tunnels, pumping stations and hydroelectric works. Lesotho will benefit from the additional development of new roads, telecommunications, health clinics, community-based rural development projects, and tourism. Earnings from the sale of its water will also constitute a major part of Lesotho's export earnings and provide much needed income to aid further development. The total value of the projects in the Lesotho Highlands Region is in excess of £2 billion. The first phase, with a value of £84 million, opened in 1998.

The possibility of transferring water in this way was first recognised in the 1950s. A number of reports were prepared, culminating in a feasibility study begun in 1982 and completed in 1986. The feasibility study was carried out jointly on behalf of the governments of Lesotho and the RSA.

The scope of the works and its cross-boundary effects involved the formation of a treaty between Lesotho and RSA. This treaty was signed in 1986, and finance was secured with assistance from the World Bank.

5.4.2. Control of environmental and social effects

It was accepted that such projects bring disadvantages as well as benefits. Local communities would be displaced by inundating land, and an influx of foreign workers and job seekers could damage local culture and traditions together with the negative effects of many other environmental and social impacts. It was, however, recognised that the positive benefits would outweigh the drawbacks, and provisions were made in the treaty stipulating that the standard of living of the affected population should not be inferior to that existing prior to project implementation.

In order to plan and monitor these requirements, social and environmental considerations were highlighted and given their own place at a high level in the management structure of the project (see Fig. 5.1).

The Lesotho Highlands Development Agency (LHDA) was formed under the control of the Ministry of National Resources of Lesotho. LDHA acts as the owner of the project, engages engineers and contractors and coordinates the project with government ministries and public corporations.

Plate 1. Katse dam under construction in the highlands of Lesotho (courtesy Institution of Civil Engineers)

Plate 2. Rural village in Lesotho now overlooked by the Katse dam access road (courtesy Institution of Civil Engineers)

Figure 5.1 illustrates the structure of the LHDA organisation which manages the project. It can be seen that executive committees were created to control finance, engineering and construction, operations and environment, and public affairs.

No specific national environmental guidelines or procedures were in operation in Lesotho prior to the start of this project. The LHDA therefore adopted guidelines and detailed environmental specifications from other countries and international agencies, including the World Bank operational directives.

> These specifications covered aspects within the control of the contractors such as water discharge from the works, protection of flora and fauna, spoil dumps, dust control, landscaping and grassing and general rehabilitation. Further obligations for health and safety,

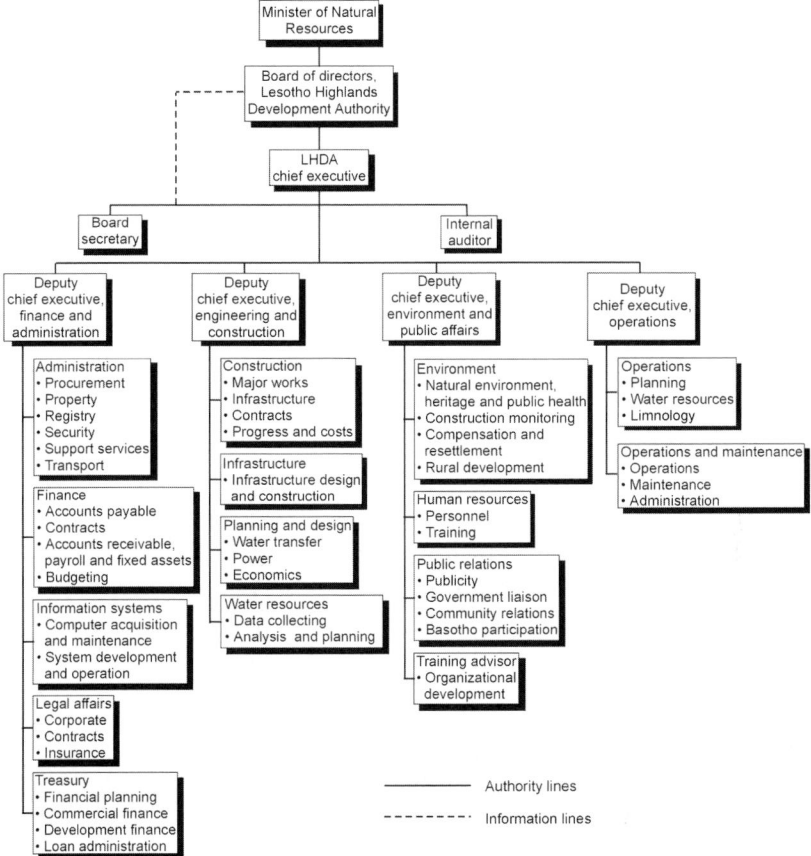

Figure 5.1. Lesotho Highlands Development Authority organization

provision of access for local communities to the project health facilities and local community relations were also specified.

The LHDA commissioned a series of social and environmental studies and developed an environmental action plan concerning compensation, natural environment and heritage, public health and rural development. Guidelines were drawn up for those communities affected by the project in terms of loss of land, income or resettlement.

5.4.3. Conclusion

This project, which is described in an excellent paper in the *Proceedings of the Institution of Civil Engineers*[1], illustrates how the management organisation of the planning and construction of a major project can and should include provision for social and environmental aspects. It can be seen that by having such a structure in place the organisation can investigate the consequences of the project on the local population and ensure that problems which do arise are dealt with on a pre-planned rather than an inadequate ad hoc basis.

5.5. Channel Tunnel — the conflict between rival 'goods'

5.5.1. Introduction – the nature of the project

This project has been described as the project of the century. It represents an investment of about £9000 million. It generated about 250 000 man-years of work, and occupies about 500 acres of land in East Kent, and a slightly greater area in France. It took nearly seven years to construct. It required a high degree of international cooperation to ensure its completion. It is used here to illustrate some aspects of the methods by which conflicts with public and private interests were reduced by formal consultation and deliberation.

The social consequences in East Kent were considerable. Apart from the influx of construction workers, a large number of additional secondary jobs were generated by the increased traffic and the resulting demand for trade and service industries. The existing social services had to be reviewed to ensure that they adjusted to the changing pattern of work and traffic in the area. All these factors and many others were taken into account in the deliberations of the parliamentary committees conducting the public investigation into the consequences of the project.

The experience of the facilities themselves by both users and neighbours was of great importance. The impact of additional traffic was accommodated by the use of the existing motorway, which had ample spare capacity. Traffic was discouraged from using the minor roads by the use of appropriate signing. The noise effects were reduced by carefully placed mounding and planting, and by the use of specially constructed screen walls. The road and site lighting were the subjects of special research.

---

[1]   Nthako S. and Griffiths A.L. Lesotho Highlands water project — project management. *Proc. Instn. Civ. Engrs, Civ. Engng, Lesotho Highlands water project*, 1997, 3–13.

The latest developments in high mast and directional lighting were investigated to reduce the amount of night-time glare evident to the neighbouring communities.

The scheme is of major significance in the development of the European Community. This was perhaps best illustrated by the cover of a prospectus for Eurotunnel, showing Europe as a series of interconnecting cogs. The Tunnel was the latest cog, the turning of which set the whole of Europe in motion.

5.5.2. Environmental context

In the UK, there are areas specially dedicated to the environment, defined as Areas of High Landscape Value (HLV), or areas of High Nature Conservancy Value (HNVC), or of Special Scientific Interest (SSI). The Folkestone terminal site encroached on such areas. It was

Plate 3. Channel Tunnel under construction (courtesy QA Photos Ltd, Folkestone)

particularly important to identify clearly the special features of the site, and to make provision to ameliorate the effects of the development, as far as possible. In addition to the care needed to minimise the impact of the scheme on the environment, it was necessary to consider the effect on local services.

The impact at Shakespeare Cliffs, to the south of Dover, is much less extensive, but is still very significant. The chalk cliffs of England have a particular emotional significance for the inhabitants of not only East Kent, but of England as a whole.

5.5.3. Range of influence
Any major project has a significant impact on a wide variety of interested parties. In deciding whether or not to proceed with the development, or in what form it shall proceed, these interests need to be weighed

Plate 4. Channel Tunnel under construction (courtesy QA Photos Ltd, Folkestone)

against each other. If the project has an impact on the national scale, environmentally, socially, economically or politically, perhaps with international implications also, then the decision making is vested in Parliament. But in a democratic nation, Parliament has a duty to hear the views of the parties concerned, and to call on the best professional advice available to help it to make up its mind.

If the application of our scale of influences shown in Fig. 2.6 is considered relative to this major project, it can be seen that there are conflicting 'goods' evident in the interests of the passive inhabitants in the vicinity of the onshore facilities, and the active users of those facilities. These needed to be weighed against each other and this was the work of Parliament. Not everyone could be satisfied with the decisions reached. Both individuals and families had to be considered in the context of the national community. Many have benefited, but some have suffered.

The wider effects on the community and the nation as a whole were appreciated. These ranged from the benefits of increased local trade to the development of much stronger trading links with the rest of Europe. The Tunnel reduces the risks of interference from adverse weather conditions as well as providing the necessary increased capacity for cross-Channel traffic.

It may be that this project had few effects on the state of the world, or of mankind as a whole — unless a stronger and more confident Europe is a stabilising and valuable part of the world community.

5.5.4. The enabling legislation

During the select committee procedure stages in the passage of the enabling legislation, consideration was given to the conflicting interests involved, not only by the committee itself, but also by the petitioners and the experts. The maintenance of a balance between the legislators, the experts, and the petitioners is of particular interest to the professional engineer who may be called as an expert. The several contributions made to the debate, representing potentially conflicting interests, are discussed below.

5.5.5. The Promoters

The Channel Tunnel Bill was a hybrid bill, jointly promoted by the developers and by the Department of Transport, the Minister concerned acting as the Sponsor in the House. The Government was thus a party to the development, but the legislation had to be presented to, and approved by, Parliament in the usual manner. The promoting department was thus an interested party, and could not act as an independent arbitrator. Its expertise and experience supported the project.

The private sector party to the Channel Tunnel Bill represents the commercial interests of the investors and operators of the project. The technical and professional skills are provided by the Anglo-French contractors with their various subcontractors and consultants. All these parties have vested interests in securing an efficient and economically viable project.

5.5.6. The Legislators

The elected, selected and hereditary members of both Houses of Parliament are concerned with the widest aspects of the project, but are also concerned with the individual interests of their constituents. They may have a particular political stance based on the policies of their party, on environmental or employment issues, for example. The range of their interests, however, gives some possibility of general issues being canvassed and supported, as distinct from the particular interests of the promoters. The members of the House of Lords bring a more individual expertise and experience to the debate which may have less political bias than is the case with elected members of the House of Commons. The informed and intelligent lay questioning by the parliamentary committees plays a vital role in ensuring that the final amended Bill takes into consideration the interests of the petitioners and the other parties.

5.5.7. The Petitioners

Any group or individual whose interests are felt to be affected by the legislation, and whose 'locus standi' gives him the right to be heard, can petition to have his disadvantages redressed by appropriate amendments to the Bill, or by legally binding assurances from the promoters. In this case the petitioners varied from the major local authorities and the ferry operators, to the local sports clubs and private individuals owning land, or whose livelihood might be affected.

Petitions cannot be accepted against the principle of the legislation, which remains the prerogative of Parliament. Petitioners have the right to cross-examine the promoters and their expert witnesses, where these are called to give evidence, or to call upon expert opinion themselves. There were several thousand individual petitions, the committee grouped these into a smaller number of topics, combining similar petitions for the purpose of expediting the proceedings.

The select committee is formed solely to deal with the petitions, to hear evidence, and to report to the House on any amendments that may be appropriate in order to reduce the adverse consequences of the project on the interested parties. The interests of the petitioners form one of the most significant aspects of this form of legislation.

5.5.8. *The Experts*

The final group of parties to take part in the consideration of the legislation are those professional persons whose training and experience provide them with the knowledge to advise the committee, and, through the committee, Parliament, on the many technical areas of special interest. These areas might include, archaeology, ecology, economics, employment, engineering, industry, land use, maritime affairs, social services, transportation, etc. Their advice and comments have a significant effect upon the decisions reached, and have considerable ethical consequences.

While the expert will be retained by one or other of the interested parties, it is his duty to advise the committee objectively from his personal knowledge and skill. It is well known that experts differ in their judgements from time to time, and it is important to assist the committee in distinguishing between the facts and the opinions laid before it. Subjected to vigorous cross-examination by counsel advocating the interests of the various parties, it is not always easy to retain a high degree of impartiality, integrity and independence. Twelve-hour days and sometimes nights, with conferences before and after committee hearings are not uncommon. The author recalls being reprimanded by his legal counsel for being too rhetorical in his evidence — 'You just give me the facts, leave me to put in the rhetoric!'

It is possible to sum up the interests of the various parties briefly as follows.

- The promoters have the commercial interests of the private investors in mind, and also those of government policy as expressed through the sponsoring department.
- Parliament has in mind the best interests of the nation as a whole, as seen through the policies of the several political parties. It also cares for the individual rights of the electorate.
- The petitioners clearly have in mind their own self-interests, or the interests of their particular organisation.
- The experts have in mind the factual aspects of the project, and the likely consequences of such developments on areas within their professional competence. It is particularly important that such advice

be given impartially, competently, responsibly, and independently, with proper professional integrity and discretion, perhaps especially by the lawyers, who frequently play a crucial role in orchestrating the activities of all the other parties.

While the procedures may seem to be elaborate, and perhaps over-bureaucratic, the systematic approach to conflict examination and resolution reduces the effect of emotional personal opinions and contributes to the sense of 'fairness' if goodwill and trust are to be maintained in society, and if objective professional opinions and judgements are to make their full contribution to the general well-being of the community. The implicit developed wisdom of the system is that conflicts are better avoided rather than being unfairly determined by superior commercial or political powers.

6

The project ethical audit

6.1. Introduction

What does ethical behaviour mean in practice? How are we to use rules of practice and codes of behaviour?

In dealing with the responsibilities and meeting the ethical concerns at each stage in a project, the professional has to be able to handle the conflicts that may arise within and between the stages, and between the interests of the several parties involved. Careful proactive planning will ensure that the consideration of potential conflicts is part of the development process, taking place, as far as is possible, before the conflicts arise. It is important for the engineer to be aware of the values held by all the different parties in any situation and negotiate a general understanding of their several responsibilities and interests. Ethics, considered as the science of human relationships, is an examination of the quality of life of societies and of the impact of proposed actions upon this quality. Due consideration enables sound, just exchanges to take place. Unethical behaviour leads to conflicts, to the misuse of resources — human and material — to poor and careless workmanship and management, and to a lower quality of life for all affected.

Ethical behaviour, based on the developed intrinsic virtues discussed in Section 3.2. (integrity, independence, impartiality, responsibility, competence and discretion), and guided by the extrinsic directions of professional organisations, leads to the exercise of competence, efficiency and collaboration. These result in good workmanship and management, and to a higher quality of life for both those concerned with the execution of a project and those affected by it.

The three levels of consequence discussed in Section 3.5.1 as of concern to the professional — general, societal and professional — can all be affected by unethical behaviour. A limited viewpoint can reduce and impair our judgement, diminish our awareness of local factors or needs, leading to partial considerations of the consequences, causing lasting deleterious effects upon the environment or upon local populations. Such actions are addressed by the ICE Rules 1 and 3.

Relationships with clients, other professionals, contractors, suppliers and official bodies must be open and fully understood and documented if conflicts of interest are to be avoided or identified and resolved. These issues are covered by Rules 4, 5, 11, 12 and 13.

The ability of professionals to discharge their responsibilities competently depends upon their personal knowledge and standards and upon the trust they have from those they serve and with whom they collaborate. Directions covering these areas of concern are given in Rules 2, 6, 7, 8, 9, 14, 15 and 16.

To enable a systematic approach to be adopted to the resolution of questions which are essentially unquantifiable, it is useful to consider the normal stages of project realisation and to review them against the professional rules of practice. Effectively, this involves carrying out an ethical audit of the project so that essential decisions can be made which include, in addition to the normal technical, economic and legal components, the ethical dimensions.

6.1.1. Aspects of project realisation

Five aspects of project realisation have been illustrated in Fig. 6.1. They are

- the interests and roles of the several decision makers
- the brief
- the context of realisation — political, social, economic, physical
- the design
- the implementation.

Each of these aspects presents a set of decisions which have ethical implications. Reflections upon these, using the ICE Rules, will identify the areas where conflicts may arise, and where consensus decisions may need to be negotiated.

The best way to resolve conflicts is to avoid them in the first place. Good behaviour helps. The engineer must develop an intrinsic set of values and work from them in order to appreciate and use the ICE Rules. Conflicts are symptoms of differences — i.e. of differences in understanding about

1. The decision makers

Notes on parties involved — decision makers, advisers, other parties

Interdisciplinary teamworking

Authority — extrinsic/intrinsic

Motives, aims, objects

Conflicts — divergent, convergent negotiations

2. The brief

Wants

Needs

Constraints

3. The context

Societal context

Physical context

Constraints

4. The project design

Long term effects — designer responsibilities

Reversibility — designer/owner responsibilities

Sustainability — all responsible

Maintenance — owner responsibilities

Environmental impact assessment

5. Implementation

Team structure and procedures

Control systems — monitoring, quality assurance, CDM regulations

Construction issues

Programme

Labour and material policies

Equipment

Figure 6.1. The project ethical audit

intentions, of differing value systems, of differing cultures, from confusions between the aims of an action and the objects of the actors. All actions in which conflicts are generated arise from our relationships within communities of people, societies, collaborative ventures. The structure and traditions of particular societies affect our moral judgements. The resolution of conflict results in agreement. Agreement requires an acceptance of a common aim, rather than differing objectives.

At each stage in the hierarchy of decisions an increasingly large number of people in an ever-widening context are affected. It is important to appreciate the ranges of influence of decisions and therefore the range of understanding and objectivity required if decisions are to be 'good'. This appreciation should be present at the earliest possible stage in the development of a project. A 'good' decision is one that improves the quality of the output and of the lives of those concerned with its execution and its result.

In general, the areas of concern in civil engineering projects are those of society, humanity and the environment. Irresponsible performance can have adverse effects on the quality of life of many people not directly concerned with the project, and on the environment, certainly for many decades if not for centuries to come. Conflicts at this level may become evident during public consultation or at formal planning inquiries, and should certainly form part of any audit.

6.1.2. Ethical audit procedures

There are clear differences between the contractual obligations entered into by the several parties to a project, and the ethical duties of those parties. The project brief should set out the aims and constraints of the project, and the duties and obligations of the several parties, acknowledging the rewards and methods of payment for the services provided, and the quality of service and product expected. The temporary relationships, formal and informal, between the contracting parties during the lifetime of a particular project require careful consideration.

A review of each contract and of the total project brief and programme is necessary. A careful selection of the appropriate person to carry out each task, using the levels of discrimination and the schedule of qualities listed in Chapters 2 and 3, will help to reduce conflict. The ICE Rules of Practice have been formulated to address the three different levels of concern — general, societal and professional — and form a valuable focus for the preparation of an audit.

The inclusion of an ethical section in every project brief will help considerably to reduce the conflicts that inevitably arise under the pressure of project execution. Such ethical reviews should be made a formal part of project procedures — with time set aside to consider relationships, or any changes needed as the work proceeds. It is best that this review be independent of normal project team meetings, and not just included as a late item on an already crowded agenda. The time allocated for such reviews will be found to pay for itself many times over. It is certainly not just a public relations item. With a clear statement of values each of the parties can carry out his individual duties with responsibility.

The five stages of a project set out in Fig. 6.1 are considered in detail below. The ethical aspects of the decision making process are explained, with reference to the new ICE Rules of Practice. In each of the accompanying tables of factors, those items which are most likely to encourage integrated teamworking, and those in which conflicts are most likely to arise, are listed.

A practical illustration of these factors is included in Appendix 1. The procurement of the Mount Pleasant Airport facilities in the Falkland Islands has been chosen since it embodies clear examples of many common problem areas. It also contains some unusual, but informative, decisions.

6.2. The decision makers

6.2.1. Introduction

The avoidances of disputes and conflicts requires careful and willing collaboration between the parties involved, at the outset of any collaborative effort. There must be a clear understanding of the aim of the project. What need is being met, and what are the surrounding circumstances. Physical, financial and temporal restraints need to be clearly defined, and the nature of the contribution required from each member of the team, established and recorded as understood. It should be acknowledged that the individual parties may have differing objectives and these should be clearly and openly defined.

Clients of the construction industry are widely varied in size, function and organisational type. They may be individuals, corporate or public bodies. Each will have a distinct philosophy governing its objectives and the way that it operates. Their priorities are likely to be radically different from those of their professional advisers.

Clients may be experienced in the construction process or they may be inexperienced, having no previous knowledge of construction. A further distinction may also be made between clients whose main business involves construction projects — developers, industrial/manufacturing organisations, public sector agencies — and those whose involvement with construction is only an occasional event (extensions to their business properties).

The client will require a construction which is fit for purpose, and one that is delivered to time and within an agreed budget. The consultant may wish to have a project which demonstrates his skill and can be provided within his fees. The contractor will prefer a project which is easily constructed and managed, and free from amendments during the construction period. The approving authority requires a project which accords with by-laws and planning rules. These are all desirable objects, but can lead to conflicts, not from dishonesty or incompetence, but from differing objectives. Hence the importance of readily agreed aims.

Table 6.1 sets out the differing values and objectives of the parties to any project. The professional engineer may find himself working for any of these groups, tasked with achieving their objects within an overall aim. The main parties influencing the quality and nature of a project are

- the client
- the consultants
- the contractors and suppliers
- the public agencies — planners, local authorities, health and safety groups, etc.
- employees of all parties — trades unions
- special interest groups — environmental pressure groups, recreational interests.

In considering these parties during an ethical audit, reference is made to the Rules of Practice of the ICE. Specific clauses are noted in the tables and text and may be referred to by using the summary of the Rules given in Chapter 3, Table 3.4. Those issues with an ethical content are noted in Table 6.1. A 'start-up' audit manual for any project would acknowledge these and report on the significance of each, amending the manual as the nature of the project developed (see Chapter 7).

Somewhat paradoxically this closer integration of interests is taking place in parallel with the move from 'status to contract'. Professional consultants may find themselves operating as design contractors or subcontractors, or as servants of their employers, rather than as respected agents of their clients.

Successful projects are now seen to be based on a number of factors.

- A clear definition of the client's brief at the outset. This is of paramount importance because without a clear aim there is little to guide the project team towards a common goal.
- Close working relationships between client and contractor from the outset, with genuine commitment from senior personnel of both groups.
- Adequate time to develop the client's brief and to carry out feasibility studies or value engineering exercises.

Table 6.1 (pages 97–99). The project ethical audit: the decision makers

Factors	ICE Rule references	Ethical considerations
Client's interests		
Maximum return on investment	1	Importance of long-term quality in public/user interests
Minimum cost	3	False economies
Best value for money		Exploitation of staff
Interests of shareholders		Definition/delegation of responsibility
Retention of good staff	**6, 7**	
Good public relationships	**3, 11**	Independent/impartial judgement
Reliability of construction team	5	
Consultant's interests		
Clear instructions	**2, 3, 5, 11**	Clarity of definition of brief, avoidance of covert obligations, evasions of duties
Acceptance as client's agent	**1, 8, 15, 16**	
Prestige	**1, 2**	Undue publicity, claiming credit due to others
Adequate fee arrangements	**4, 5, 13, 14**	Exploitation of staff, ensuring competence
An interesting project	12	
Staff opportunities	**6, 7, 8**	
Competent contractors	5	Bias in selection of contractors and suppliers
Contractors' and suppliers' interests		
Minimal alterations	**1, 3, 11**	Health and safety issues.
Enhanced reputation	**2, 7**	Selection of suppliers/ subcontractors
Interests of workforce	**6, 7**	
Minimise costs	14	Unfair competition
Compensation for extra work	5	
Maximum profit	**5, 11**	
Negotiation of costs of variations		
Maintenance of quality		
Adequacy of staff, competence of workforce, care of staff and workforce		

Table 6.1 continued

Factors	ICE Rule references	Ethical considerations
Public agent's interests		
By-laws	4	Application of rules in total project context
Compliance	4	Relationships between codes and professional judgement
Ease of assessment	2	
Health and safety	3	Clarity of relationships between components and total project
Planning gain	3, 11	
Special interest groups		
Opinion formation	1, 2, 4, 8	Demonstration of appropriateness of ends and means
Local employment opportunities	1, 3, 5, 11	
Minimal environmental impact	3, 4, 11	Discrimination between opinions and objective judgements
Sustainability	3	
Amenity/sport/recreation	3	Avoidance of conflicts of interest
Employees' interests		
Career progression	1, 3	Balance between individual development, training, and contractual duties
Job satisfaction	2, 7	
Working conditions	2	Opportunities for responsibility, but avoiding risk-taking
Pay	6, 7	
	8, 14	Good working conditions, avoidance of over-demanding work schedules
	16	Adequacy of rewards

Table 6.1 continued

Factors	ICE Rule references	Ethical considerations
Integrating factors		
Common goals/trade-offs —	1, 5, 11	Acknowledgement of all team members' duties. Definition of members' authorities
Reputation/satisfaction/profit	2	
Team working	7	
Clarity of brief	1, 5, 11	Agreement on total and sectoral responsibilities and scope of work
Definition of responsibilities		
Communication/negotiating skills	7	Communication procedures between all parties
Transparency in transactions	11	Agreement on major areas of risk
Mutual trust	1, 2	
Recognition of particular contributions	7	
Extrinsic/intrinsic authority	1, 4, 12	
Risk and security control	3	
Conflicting factors		
Divergent interests	1, 3, 5, 11	Conflicts of interests between parties
Unjust balance of work and rewards	2, 8, 13–16	Equity between parties
Secrecy	4, 11	Possibility of covert agreements between some parties
Incompetence	7	
Lack of recognition of authority	1, 2, 4, 5	Undue assumptions of authority
Risk exposure	3	Conflicts between financial, programme and human interests
Lack of consideration for workforce	2, 6–10, 14–16	

- The achievement of project aims requires that all the team-members agree on the principles from which they are working, have a good understanding of the context within which they will be operating, and are aware of the resources available to them. The objectives of the various team-members can then be harmonised to meet this agreed aim.

- Early agreement is required on broad parameters for the commercial viability of the project.
- Flexible approaches are required to incorporating client changes during the design stage.
- 'Open-book' approach to negotiations for prices of construction work is required, particularly for estimates of the cost of variations.
- Agreement on detailed specifications is required for projects before the start of construction works.
- Consideration must be given to the matter of standardisation, to ease buildability.
- The use of independent professional project management, using experienced professionals, should be considered.
- Structured total quality management (TQM)/quality assurance (QA) should be utilised.
- Good working relationships between parties and good communications are paramount.[1]

6.2.2. The variety of interests

The nature of the changes set out in Section 6.2.1., to the traditionally accepted construction philosophy has inevitably led to a redefinition of the duties, roles and responsibilities of the engineer. The roles may be less clearly defined and may generate some internal conflicts. However, conflicts arise under the traditional and the clearly defined roles of independent organisations of specialists — clients, consultants, suppliers, contractors, financiers, etc.

In considering these diverse interests, it is helpful to take into account those factors which assist the reaching of common aims and standards — the integrating factors, and those conflicting factors which arise from differences. Examples of these are given in Table 6.1.

At the outset of any project the client begins to assemble, probably with the assistance of his prime consultant, the team of advisers, suppliers and contractors with the necessary skills to achieve his objectives. The following factors relate to the development of such a team.

- It is unlikely that any one adviser will have the skills and knowledge necessary to develop a project. Briefing becomes a team effort as professionals with specific expertise or experience are introduced to

---

[1]   Bedalian, H.M. Summarised from 'Successful major projects'. *Proc. Instn Civ. Engrs Civ. Engng*, 1996, **114**, Aug., 117–123.

the project. Many of these will form part of a permanent project team, others may be retained only as long as their input is required.

- The project and design teams may comprise many specialist disciplines, civil, structural, mechanical and electrical engineers, architects, land surveyors, quantity surveyors, economists, etc. depending on the scope and magnitude of the project.
- The number of parties involved in a project may well reach a level where a project co-ordinator or project manager will be formally appointed. This is frequently the case.
- As the aims and means of achieving them, in a project are established and developed, the scheme is subject to appraisal as to its viability. Sketch designs will be prepared by the design team and a strategic programme prepared reflecting the clients aims, completion date, financial targets, and quality standards.
- Each member of the team, including the client, will be faced with a variety of problems and each will offer solutions based upon their background, experience and professional training. The project manager will focus the analyses of these problems in the context of the overall project goals, presenting the alternatives to the client for approval.
- These discussions will involve or be influenced by the views and concerns of external organisations and interested parties, many of whom may be affected by the project during its planning, construction or operational stages, and in the long or short-term. Their concerns may include social considerations such as employment, health and safety, education, cultural or recreational activities, security, displacement and relocation. They may also be interested in the environmental or other aspects of the project.

In all these situations the engineer, who may be playing any one of several roles — client, consultant, public agent, contractor, etc. — must assess his own duties and responsibilities relative to those roles. The ethical audit will identify the roles of the various parties, and attempt to define their responsibilities within the total procurement process.

6.3. Formulation of the brief

6.3.1. Introduction

All project designs are the result of attempting to meet a particular client's need within a particular context, with the particular abilities and resources of a construction team. In this Section, the preparation of a client's brief for a project is discussed. The preparation of the brief requires careful consideration not only in order to meet the needs appropriately and efficiently, but to avoid ethical conflicts arising from partial views, or misunderstandings.

At the inception stage of any project the client perceives a need. This will be a response to the situation in which his organisation operates. An opportunity to be exploited. Generally, the need will be a combination of many forces and stimuli including the need to survive, to maintain position in relation to competitors, economic forces, sociological forces or needs. These needs trigger the start of a construction project — although construction professionals may not be involved initially. At this time the client probably has little information on what he requires from the project, other than a broad estimate of how much money is available for the works and an approximate idea of the time at which he requires the project to be completed and operating.

For many clients approaching a construction project there will be feelings of vulnerability based upon

- working outside their own field of expertise or knowledge
- placing themselves in the hands of professionals who may not understand their business objectives or methods, and who will be using working methods and technologies that the client may not understand or have trouble in controlling
- uncertainty as to the feasibility or programme of the project
- knowledge that they may be embarking on a complex process with high risks.

The client will require 'value for money' measured in terms of time, quality and price. The relative importance of these three parameters will vary from client to client and each will be willing to make 'trade-offs' between the three in order to achieve his own particular organisation's goals.

The client will require the assistance of professional advisers to help him to appraise the feasibility and viability of his plans and ultimately to assist him to construct the project, should he decide to go ahead. He will aim to pass some of the risks involved in the project onto his advisers and the constructor. He will expect the advisers to provide sound independent advice and information in a form that he can understand and which can be used to help him to make decisions. He also expects his advisers to understand his needs, context, and operational aims and structure so that their advice will be relevant. The focus of these needs may be internal to his organisation and may conflict with the broader outward looking views of his advisers.

6.3.2. Formulation of the brief

Projects of all sizes involve the appraisal and consideration of a multitude of constraints — technical, legal, financial, environmental, organisational and ethical. The client's brief will ultimately define the priorities and objectives of the project. Without a comprehensive brief no project can be successful.

The brief is usually prepared in an interactive dialogue between the client and his adviser, with the adviser obtaining and clarifying information from the client. Part of this information may be obvious and easy to access. Much may be unclear and the client may not appreciate its relevance to the adviser in the context of the project. This may be more clearly understood by reference to a 'JOHARI' window diagram — after Bejder (1991).[1]

The 'panes' in the window diagram of Fig. 6.2 can be used to illustrate the briefing process. At the time of the briefing process, client and engineer attempt to define all aspects of the project so that the full significance and implications of the work can be studied, probably as part of a feasibility study. The diagram shows the process of communication. The client and the adviser are trying to 'push back the dotted line' to reveal and to explore all the requirements of the project. The diagram thereby demonstrates the importance of disclosure and feedback.

At this time both client and engineer have different perceptions of the needs and scope of the works. A common level of understanding must be reached as quickly as possible if their 'partnership' and project are to succeed without conflict. From Fig. 6.2 it can be seen that there are four areas of interaction.

Known to client

This shows the areas of the project where the client has information or requirements that can be clearly identified at briefing stage. Both parties immediately see the implications of this information.

Unknown to client

This represents the area where the client may be unable to see or understand the full implications, or the consequences, of his requirements. The adviser,

---

[1] Bejder E. (Barnett P.S. and Miles E.R. (eds)). From client's brief to end use: the pursuit of quality. *Practice management: New perspectives for the construction professional*. 1991, 193–203.

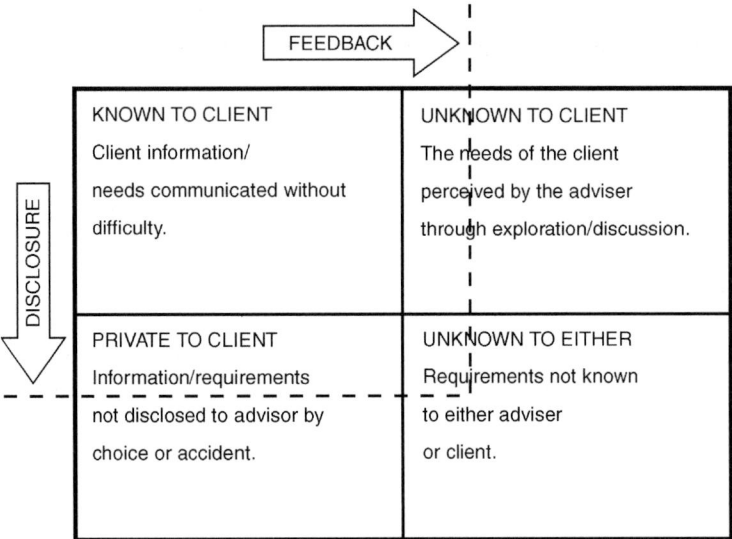

Figure 6.2. JOHARI Window Diagram showing briefing process

often an engineer, because of his experience and knowledge, is aware of these elements. The adviser is then able to explain the effects of these requirements and incorporate them into the project study.

Private to client
This represents areas of the client's knowledge or interests which he is unwilling to disclose to his advisers. These may be process or financial considerations which, while they have a bearing on the project, cannot be revealed initially. It is frequently important that such 'private' information is revealed because of its potential influence upon the project. The need here is for the client to have trust in the discretion of his advisers.

Unknown to either
This represents the aspects or consequences of the project of which neither client nor adviser is aware at the beginning of the briefing stage. These aspects may become apparent during the development of the brief, or they may only be revealed as the project is developed, constructed or even commissioned and operated.

Figure 6.2 shows the parties involved in the briefing process attempting to extend the boundaries of their common knowledge. The experience and knowledge of the adviser may greatly facilitate this process. An experienced client will see more of the 'window' at the outset, recognise the importance of disclosure and greatly improve the briefing process. The development of trust between the parties involved is essential because it helps the client to reveal more of the private information at his disposal and be more receptive to the ideas of the adviser. The unknown area may be uncovered as a result of both client and adviser developing a new view of the project, as a result of their co-operation.

The briefing process takes into account not just technical requirements but all social, environmental, financial and ethical aspects of the project which must be considered if the project is to be successful. Table 6.2 gives some guidance as to the ethical issues which may arise during this process, with references to the ICE Rules of Practice relevant to each. Key issues are the clarity of responsibilities, the importance of trust between all parties together with the associated implications of trustworthiness.

6.4. Context

6.4.1. General issues

The nature of society was considered in Chapter 2. Before we can make any firm decisions for ourselves on what is 'good' and what is 'right' we need to have an understanding both of societies and of mankind, and of the context within which we operate and make decisions.

Individuals living together in societies improve the quality of their lives by their corporate efforts and their shared skills and talents. That such efforts frequently lead to misery and distress is evident, but this may be more often the result of ignorance and greed than a reflection of the nature of societies in general.

Engineering projects inevitably impact not only on the physical, but also on the economic and social 'environments' that contain them. 'Environments' include all the factors outside the project that may affect, or be affected by, the project, not only the 'green' issues. These impacts may be beneficial or they may be harmful.

Modern engineers, whilst still accepting the 19th century role of civil engineering being, 'the art of directing the great sources of power in nature for the use and convenience of man', must now also take into account finite resources and the complex interaction of the project and its environment — social and physical.

Engineers are therefore responsible not only for the effective use, management and conservation of the resources placed at their disposal within the project, but also for the analysis of the project within a complex and changing external environment. In this modern role the engineer

Table 6.2. The project ethical audit: the brief

Factors	ICE Rule references	Ethical considerations
Economic factors		
Return on investment	**3, 5**	Value for money
Low prime cost		Risk
Low running costs — energy		Prime cost v. revenue costs
balance		Energy consumption/audit
Cash flow		Sustainability
Space		
Adequate space standards	**3, 7, 11**	Clarity of definition
Arrangements of space —		Quality of life of users
gross/net space ratio		Amenity
Design standards	**4**	Any special cultural require-
Preferred materials		ments
Maintenance management/		Non-toxic, low-energy,
costs		recyclable materials
Public space		Space between buildings
Facilities, equipment, services		
The site		
Environmental factors	**4, 5, 11**	Environmental impact
Alternative locations	**2, 7**	assessment
Values — developed,		Land use, economy
undeveloped		
Legal factors		
Planning conditions	**1, 2, 4, 12**	Make restrictions clear
Responsibilities	**16**	Define responsibilities
Liabilities		Define decision making roles

participates in a system that protects communities and the natural environment from the adverse effects of the project and helps to ensure that free market forces alone do not decide the development of a project.

Table 6.2 continued

Factors	ICE Rule references	Ethical considerations
Contractual procedures		
Competitive bids	**4, 5, 11, 13**	Clarity of liability
Manage/design/build	**7, 8, 14**	Fair competition
Private finance		Quality management and control
Insurances		Security of documents/ information
Programme		Conflicts of interest
		Adequacy of contract period
Integrating factors		
Clarity of definition of brief	**1, 2, 5, 7**	Statement of project aims/objectives
Itemisation of elements		Careful documentation/ definition
Mutual understanding		Clarity of responsibilities
		Equity of rewards
Conflicting factors		
Confusion of meaning — assumptions of understanding	**7, 11**	Poor definition of key terms
Confusion of wants and needs		Intrusion of particular individual preferences/ preconceptions
		Negligence

The engineer must work in a system of formal regulations, institutional rules and personal ethics. Government exercises wide-ranging powers of control over the planning and construction process. It has the role to review and take account of the conflicting demands of industry, commerce, public health and safety, transport policy, employment and many other aspects of a modern complex society. To achieve these ends, government provides a broad range of statutory regulations, controlling development planning and environmental issues, together with legislation covering health and safety, labour, security, traffic, etc.

Such controls significantly influence the environment in which projects are planned, constructed and operated, as do the personal value systems of the parties concerned and the professional codes of the several regulating institutions.

6.4.2. The environment

Concern for the care of the environment has increased considerably during the course of recent decades, in parallel with our increased understanding of the consequences of our actions. It is necessary to give special attention to this aspect of the work of the engineer, although it is to be recognised as being only one of several issues to be addressed.

The third Rule in the new Code states the following.

Rule 3. A member shall have full regard for the public interest, particularly in relation to the environment and to matters of health and safety.

Comment
He should ensure that systematic reviews of all aspects of the project's impact on the environment are taken to diminish any adverse effects. Due regard must be given to health and safety, both in the immediate and the long term.

The engineer is therefore committed to a concern for environmental protection and even improvement.

> If a clean environment is promoted to protect human health, how do we measure 'clean'?
> If a project includes the construction of a dam that will destroy a 'natural' river and flood farmland, how should an engineer regard that destruction of a natural environment or even the viability of the overall project in terms of good or bad social policy?
> Should the engineer act and object as either his professional or social ethics demand?
> In what capacity should he object, professional or individual?
> What happens if the underlying values conflict?

An attempt to review the ethical aspects of any project involves complex analyses. Cost-benefit analyses may be almost impossible to quantify. The identification and distribution of costs and benefits may be difficult. An analysis may even produce conclusions which, while justified in engineering terms, offend the moral beliefs of 'ordinary people'. Optimising, harmonising, compromising, reconciling, all have to play their parts.

The professional engineer has the difficult task, shared with professionals in many other disciplines, of having to differentiate between his personal views and his professional duties in order to make an objective analysis of a project. Obviously, engineers must assume a responsibility for the effects of their work and endeavour to make a substantial contribution to the protection of the environment. They should not participate in projects that are unnecessarily destructive to the environment — even if those projects do not immediately endanger life or health. An engineer should be able to express publicly a professional opinion on environmental matters, but only when that opinion is based upon sound knowledge and analysis. Where his personal opinions, based on professional expertise, conflict with his client's wishes, then he may need to withdraw in order to avoid public disrespect and the generation of mistrust for the profession.

Should an engineer consider that a project is unnecessarily harmful to the environment, then he should attempt to explain the situation to the client in order that he is able to make an informed decision on the project. In the event of his opinion being disregarded, an engineer has the duty and the right to withdraw from the project, and the right to make his concerns known to the relevant authorities.

6.4.3. Interdependent factors

All projects can be viewed as part of a group of interdependent contextual factors. This is illustrated in Fig. 6.3.

Political factors

These factors represent the power of government policy. This power may influence the availability of finance, levels of taxation and incentives, together with labour and educational standards. Such policies can distort the ethical field, requiring some general 'goods' to be diluted for some particular end — tax gains, planning consents, development grants, for example. It is the duty of the professional to reduce harmful or limiting effects by maintaining a balance between rival 'goods'.

Legal factors

These factors relate to the regulations governing such aspects as safety, siting, planning, contracts between parties and associated methods of settling disputes. The importance of understanding local laws in different countries is vital in ensuring amicable and productive relationships.

Institutional factors

These factors are concerned with the influence of the professional institutions regulating training, conduct, conditions of employment, competition

Table 6.3. The project ethical audit: the context

Factors	ICE Rule references	Ethical considerations
Social context		
Employment		Cross-cultural factors
Social services	**1, 3, 4, 11**	Impact on local community
Economy		— health care, security,
Housing		education, recreation, etc.
Safety		Economic factors — property
Pollution		values, transport provision
Recreation		Availability of skilled and
Heritage		unskilled labour — working
		hours
Physical context		
Access/transport	**2, 3, 4, 11**	Environmental impact
Materials availability		assessment — see EEC
Services required (water,		directive (Table 3.3)
power)		Duration effects
Amenity		Reinstatement of land
Conservation — ecology		Shared services, e.g. drainage
Sustainability		Spoil disposal
Energy consumption		
Topology		
Duration effects		
Integrating factors		
Mutual benefits	**1, 3, 4, 11**	Good public relations,
Communications		communication skills
Thorough and public		Information flow
investigations		Adequate definition of all
Wide consultation		operations
Involvement of community		Provision for replacement/
		enhancement of social
		facilities
		Community participation

Table 6.3 continued

Factors	ICE Rule references	Ethical considerations
Conflicting factors		
Divergent interests	**3, 5, 11**	Lack of communications with local community
Inadequate information	**2, 3**	Destruction of amenities/ facilities
Covert investigations	**5, 12**	Short term aims of project team-members conflict with long-term needs of community
Change of land use	**2, 3, 4, 11**	
Site classification — SSI, HLV		Adverse impact during construction

rules, or a similar framework of issues, as given in Chapter 3, Table 3.3, is of great benefit in reducing conflict and making an objective appraisal and presentation of such issues possible. A checklist of factors involving ethical decisions is given in Table 6.4.

Key decisions will concern the balance between construction and operating costs, the availability and cost of alternative materials, and the working conditions of the users of the completed project.

Reference can be made to the hierarchy of consequences set out in Chapter 2, referring to the relative needs of society, of humanity (which could include international considerations) and of creation as a whole (including major environmental consequences, global warming, reduction in gene stock diversity, etc.). All these factors may have to be considered and agreed during the design process. The responsibilities are considerable, but the adherence to the ICE Rules will help to ensure that a proper balance is maintained, and that the decision makers are well qualified to adjudicate on such matters.

6.6. Implementation

6.6.1. General issues
There are two distinct areas of possible conflict at this stage in a project: the internal issue of relationships between the various team-members and the external issue of relationships with the affected local community.

Table 6.4. The project ethical audit: project design

Factors	ICE Rule references	Ethical considerations
Design proposals		
Alternatives	**1, 3, 4, 5, 11**	What balance of time to spend on alternatives
Optimisation		
Trade-offs	**2, 7**	Objectivity of alternative designs
Maintenance		
Materials choice		Conflicts between rival 'goods'
Relevance to context		Prime v. Operational costs
Aesthetics		Competence of design team
Impact		
Long-term effects	**3, 4, 11**	Allocation of time/resources
Reversibility		to consultation (see EEC
Sustainability		regulations — Table 3.3)
Environmental impact assessment		
Short-term impact disturbance (light/sound) pollution (effluent/dust etc.)		
Integrating factors		
Well-documented proposals	**1, 3, 4, 7, 11**	Communication systems
Objective assessment of alternatives		Demonstration of design alternatives
Objective and balanced impact assessment		Definition of impact analysis
		Identification of liaison officer
Conflicting factors		
Partial or confusing presentation	**2, 7**	Poor documentation
Prejudiced assessment of alternatives		Inaccessibility
Insensitive designs		Minimum compliance with standards
Extravagant use of materials and energy		
Short-term objectives		

If adequate consultation and communication has been developed during the earlier stages then the team members will have recognised the main aim of the project, and will have related their individual objectives to this principal aim. The factors associated with implementation are set down in Table 6.5.

6.6.2. External relationships
The interface between the project and the general public, however, usually becomes most evident at this final stage of implementation. While consultation may have taken place during the planning stages with locally interested parties, there will be many who were not involved in these discussions, and who will only have become aware of the project when physical work becomes evident. The impact of a substantial labour force and construction equipment becomes very evident, as are the additional calls made upon the local facilities — access roads, services, local labour market, etc.

To minimise the possibility of conflicts arising between local official and unofficial interests, it is important that all concerned appreciate the problems involved and that steps are taken to reduce the impact to a minimum. If the project has a considerable impact then some form of liaison group may be advisable through which concerns can be communicated, perhaps with a 'hot line' for matters of seeming immediate urgency. Explanatory talks to community groups, covering the nature of the completed development, but also explaining the construction programme and agreeing how best to reduce its impact are a valuable aid in reducing and identifying possible areas of conflict. Visits to the works by groups from the local community may help to give some sense of local 'ownership' and involvement in the project. It may also encourage collaboration with works that are almost certainly intended to improve the quality of life in the community when completed — nationally, even if not locally.

The evidence of continuous efforts to keep the environs of construction sites clean and free from pollution, to organise traffic movements in a considerate manner, recognising local interests, and to agree working hours, as well as to advise the community of any particularly noisy or extended working operations, will help to reduce conflict.

These matters are all covered by ICE Rules 1, 2, 3, 4, 7 and 11.

6.6.3. Internal relationships
In the pressure to complete projects on time and within costs, as well as providing satisfactory financial performance for the several parties

Table 6.5. The project ethical audit: implementation

Factors	ICE Rule references	Ethical considerations
Team members		
Client's role	**4, 5, 11**	Definition of areas of
Consultants	**2, 7**	responsibilities
Contractors		Agreed lines of
Suppliers		communication, external and
Public sector inspectors		internal
		Purchasing procedures
		Authority of inspectors
		Roles of site managers,
		resident engineers, local
		representatives, etc.
Control procedures		
Quality assurance	**7, 11**	
Testing procedures		
CDM Regulations		
Progress control		
Construction management		
Quality control	**3, 4, 7, 11**	Role of project manager
Materials control and supply		Need for quality management
Storage on site		officer
Plant and equipment		Independent testing and
		quality management
		Relationships between client/
		contractor
		QA role
		Contract conditions for
		subcontractors and suppliers
Labour management		
Site hygiene	**2, 3, 6, 7**	Duties and authority of site
Health and safety		safety officer
Labour relationships		
Training		
Trades union liaison		
Organisation of operative/staff training		

Table 6.5 continued

Factors	ICE Rule references	Ethical considerations
Integrating factors		
Regular all-party meetings	**1, 4, 5, 11**	Relationships of client, consultants, construction senior staff
Accurate reporting on cost and time control	**2, 7**	
Accurate and objective reporting of any contingencies		Independence Communications with local groups
Communication at all levels		
Close liaison of site management and quality control teams		Avoidance of conflict Agreed aims
Conflicting factors		
Adversarial attitudes	**1, 2, 7, 14, 16**	Partial/sectarian attitudes
Withholding of information		Promotion of limited interests
Poor site cleanliness and safety controls		Denial of responsibility Indifference to complaints
Lack of authority at construction site level		Rigid adherence to rules
Interference from client/ consultant in construction management		
Poor relationships with local community		

involved, the need for quick decision making can result in lack of consultation and conflicts of responsibility.

The organisation of the project team needs to be clearly established, with levels and areas of responsibility clearly defined before work starts. Lines of communication should be clear and maintained, access to key decision makers needs to be readily available, and senior managers must be approachable and willing to listen.

Details of the quality management organisation should be fully understood, with lines of reporting established from testing operations to line managers. Relationships between resident engineers and supervisory staff and their colleagues in the construction management team must be good, with responsibilities accepted as professional duties and not personal preferences. Well drafted contract documents contribute greatly to clarity and the anticipation of possible areas of difficulty, enabling possible conflicts to be resolved before they arise. The types of issue that can cause conflict, for example, include a need to change a material supplier, the need for extended working hours, the introduction of unplanned construction techniques — blasting, open cut substitution for tunnelling, etc.

Site hygiene and safety are the responsibility of the site management, and are part of the overall professional responsibility for all engineers involved in the project. Labour relationships require consideration and fairness. Adequate provision needs to be made for the training of all operatives and management staff, in that regard.

It can be helpful to hold initial, and perhaps regular, meetings at which the designers or owners can explain to the site operatives why certain decisions have been made, the sequence of operations, the importance of quality control and perhaps to encourage feedback from 'grass-roots' on suggestions for improvements in working procedures.

7

Summary of ethical audit procedure

7.1. Introduction

The aim of an ethical audit, as with all audits, is to establish the facts of the situation, and to compare these with the relevant plan. In this manner the actions required to maintain an effective development and delivery of a project can be taken. This should then ensure that the resulting project will be of value to the individuals concerned and affected, that it will meet the needs of the commissioning authority, will enhance the quality of life of the community, and will not cause lasting deleterious effects to the environment.

7.2. The audit statement

It is of value to set down the details of the project, not only at the outset of the development, but as they are revealed or changed during the lifetime of the work. An ethical 'file' or database can be prepared and updated regularly taking into account the several stages as given in Chapter 4. A suggested filing index is given in Table 7.1.

The continuous appraisal of the current state of the work is most important. The past is only of relevance if it informs us of present situations, or suggests possible outcomes for alternative lines of developments. The clearly established aims of the project will suggest lines of action proceeding from the known situation to the desired ends. A special meeting considering only ethical aspects of the project can be of great help. Failing this, adequate time must be set aside for their consideration during other meetings. Continuous attention to the details can reduce the time required to resolve problems to a minimum, ensuring smoother running of the contracts. From Chapter 6 of this book, and

Table 7.1. The audit statement — typical file index

File	Details
Details of parties concerned	Names and addresses, contact numbers, e-mail codes, etc.
	Decision making responsibilities and authority
	Executive officers
	Non-executive parties
	local authorities
	government departments
	special interest groups
	police
	health authorities, medical officers
Project administration structure	Organisational framework
	Reporting levels and procedures
	Finance management, control of payments.
	Claims procedures
Special operating procedures	Security
	Confidentiality
	Discipline
	Designated officers
Project details	Agreed briefs
	Identification of significant contextual issues
	Summary of key decision proposals
	Contractual restrictions and constraints
	labour, materials, transport, working hours, pollution control, access to and extent of site, trades union relationships, site security
Review meetings	Consultative meetings
	Agreed representatives to attend
	Official authorities
	Special interest groups
	Project meetings
	Hierarchy of meetings
	Attendance at meetings
	Authority of meetings
	Distribution of Minutes and Action notes
	Follow-up procedures

Table 7.1 continued

File	Details
Testing and quality management procedures	Testing authorities Reporting procedures

Contract meetings	Minutes of previous meetings
	 Reports on action arising
	 Any alterations to project design
	 Alterations to contractual procedures and programme
	 Personnel issues
	     discipline
	     changes in management responsibilities
	 Environmental issues
	     traffic
	     pollution
	     damage
	     complaints from public
	 Agreed action and identity of responsible officers

Consultative meetings	Minutes of previous meetings Report on subsequent actions Report on current state of work Comments and requests from interested parties Summary of actions/investigations agreed

from the special aspects of particular projects, standard agenda can be developed for such review meetings, with particular individuals identified to report on differing sections of the work.

7.3. The audit reviews

There are two types of review meeting — the construction team meeting, and the consultative meeting with local interested parties. Suggested agenda for these meetings are given in Table 7.2, but these will need to be considered carefully for each particular project.

The report format for the ethical audit should be in two parts, namely

- a review of the present situation
- the ethical implications of future activities.

Table 7.2. The audit review — suggested agenda for ethical review meetings

---

Contract meetings

    Minutes of previous meetings

    Reports on action arising

    Any alterations to project design

    Alterations to contractual procedures and programme

    Personnel issues

      discipline

      changes in management responsibilities

    Environmental issues

      traffic

      pollution

      damage

      complaints from public

    Agreed action and identity of responsible officers

Consultative meetings

    Minutes of previous meetings

    Report on subsequent actions

    Report on current state of work

    Comments and requests from interested parties

    Summary of actions/investigations agreed

We will consider all the five stages described in Section 6.1.1 and set out in Fig. 6.1, including a summary statement as each stage is completed and 'signed off'. There may be valuable lessons to be learnt for future projects by a review of the development of the current work, but these should be noted separately and not allowed to dominate the present concerns. The object of the audit, or ethical brief, is to identify possible points of future conflicts of interest and to eliminate them wherever possible.

The records and recommendations can be divided usefully into the above five stages.

- *Schedules of all parties concerned* — to include names, addresses, etc. management structures of each organisation, with deputies named for each key player. The roles, responsibilities and reporting paths of all the main 'actors' should be defined — clients, managers, consultants, contractors, suppliers, design/build/supply organisations, statutory officers, etc. This should include details of subcontractors, with records of their contractual relationships with main contractors.

- *The agreed briefs* — summaries of briefs for all sections of the project, including programme limitations, should be included in the ethical files. Special note must be made of any performance specifications and of any special constraints (e.g. on suppliers, country of origin, materials, safety standards, security, confidentiality, etc.). Quality management procedures need to be set out, with definitions of the responsibilities and authorities of those concerned with the monitoring of quality.
- *The context* — in addition to a full description of the site and of those aspects of the neighbouring communities likely to be affected by the work, records should include full details of any environmental impact assessment that has been carried out, with the recommendations carefully noted. Are there any special requirements, concerning working conditions, health and safety, security, pollution, communications, traffic management, restrictions on working hours, etc.?

Statutory and legal requirements should be noted, with definitions of any national or international standards to be observed. In some cases when operating between, or in, different nation states, a special section identifying any unusual aspects of the law should be noted. Relationships with approving bodies (planners, by-laws, etc.) need to be defined, and where no local regulations exist then agreement needs to be reached and recorded on standards of working practices to be observed.

- *The design* — any special requirements for reversibility, sustainability, energy conservation, etc., need to be noted. Variations in design during construction are not unknown, and the original criteria need to be available for reference, e.g. capital v. maintenance criteria and whole-life costing. The contract documents and detailed specifications and instructions issued, should clearly define and instruct contractors and suppliers on their contractual obligations. Careful attention should be given to the interface between the parts of the project, particularly if work on these is to be undertaken by different parties, perhaps separated by some interval of time.
- *Implementation* — the items set out in Section 6.6. covering relationships with local communities and the construction aspects of the

project should be noted and reviewed as a matter of routine at the regular meetings on ethical, or working practices. Trades union relationships should be noted, with reports on security and safety. Problems with the delivery and supply of materials and components, or of dust, noise and light pollution, should be resolved.

7.4. Conclusions

The identification and definition of possible sources of difference can reduce conflicts to a minimum and ensure that trust between all parties, with respect for their particular interests and contributions, underpins all the work, and results in a sound project, delivered to programme and budget, to the satisfaction of all concerned.

8

Education

> *No profession requires more education, theoretical and practical, and more training of the mind, than the profession of a civil engineer*
> Sir John Wolfe Barry, Presidential Address, 1898

8.1. Introduction

The task of providing education on ethical considerations for engineers addresses two distinct groups

- undergraduate students
- adult learners — graduate engineers seeking chartered membership together with mature engineers having practice and 'life' experience, who wish to acquire further education or continuing professional development.

Developing courses for these two groups must accommodate different learning styles but address the need to make ethics 'relevant' to professional development. Professional ethical education must prepare engineers by making them aware of the ethical dimensions arising from the environmental and socio-economic aspects of their work. These aspects must be considered when problems are investigated. A full discussion on the global context of engineering decision making and the roles of professional institutions is given in Chapters 1, 2, 3 and 6.

A system to incorporate the ethical aspects of the project in the decision making process, by carrying out an 'ethical audit' at the planning stage is detailed in Chapter 4. It is suggested that this system be used as part of the development of education in ethics. With this in mind the aims of ethical education in civil engineering can be seen as twofold.

- To enable the practitioner to think ethically. This demands the professional or pre-professional being able to practice ethical decision making as a part or dimension of the everyday professional decision making.
- To enable the practitioner to develop the skills, qualities and knowledge in order to practice ethics. This demands the practitioner taking responsibility for his own values and decision making, and for developing and integrating the necessary skills and qualities.

This Chapter will look at how these aims can be achieved in very different contexts, including professional development and undergraduate work.

8.2. Undergraduate students

8.2.1. General engineering education
Engineering undergraduate education is the process of training students to identify and solve problems through rational application of natural sciences and mathematics. The chosen solutions to these problems will almost always have consequences in human as well as engineering terms. The usual method used by students in this process is one of data gathering followed by reflection, assessment of values, etc.

Engineering students are already provided with a programme of complementary technical studies that develop a thorough grounding in materials, analysis, physical sciences and mathematics. What also must be developed is an awareness of the effects of technology on human activities and interests. This requires an understanding of the merits and limitations of technology, its perceptions by society and the potential for ethical problems that lie in these areas.

8.2.2. Methodology
Students with no professional experience to draw upon can be given a case study, as an exercise, based upon the experience of a pre-professional group, enabling them to identify with it.

8.2.3. Case study
Students from a private university in Colombia are given a project in which they are to meet some of the technological needs of a remote tribal village.

After a long and difficult journey the students reach the village and ascertain that the most urgent priority is the provision of an adequate supply of water. This would have to be derived from a large river which lies about half a mile from the village, and is some 15 m lower than it.

The crops on which the villagers depend require a regular and copious supply of water, which is not always available during the drier parts of the year — though the river continues to flow, coming from the high mountains some distance away. Furthermore, the journey on foot through mud and swamp to reach relatively clear water in the river is neither pleasant nor convenient, and has led to distinctly dubious practices with regard to the use of water for domestic purposes.

The swamp area is infested with mosquitoes and is the cause of much disease. It is also the breeding ground for a rare alligator. Further downriver this alligator is hunted for its skin, despite this practice being outlawed.

The village, of course, possesses no electricity supply, and the villagers have — at the very best — an extremely rudimentary technology. Housing conditions, in huts made of timber and willows, are extremely poor. Illness and deformity are common, and the villagers have virtually no contact with the outside world.

The culture, however, is an attractive one — in that honesty appears to be accepted as a virtue, most of the villagers are accustomed to having to work hard to keep body and soul together, and the corporate spirit of the village group is extremely strong.

It seems unlikely that anyone will attempt to support or further assist these villagers once the project is completed.

What course of action is proposed?
It is suggested that the case study should be examined by several sub-groups which are told not to find a solution but to work out a method for handling the case. The sub-groups then present their results to the whole group who test them through questions of clarification and then by disputation. The tutor can help the whole group to reflect effectively on the essential elements of method and assist in the formulation of a class methodology. This can then be compared with the methodology set out in the ethical audit in Chapter 7. It is possible to vary this approach with members of the class being invited to study and respond to a case, before the first meeting.

It is also possible to approach a 'method' by encouraging comparison with the students' normal approach to ethical decision making. The students can be invited to reflect upon their own ethical method in life in general (see Section 2.3.1).

Working on and establishing an ethical audit method provides a clear framework for reflection. It also forms the basis for any reflection on ethical theory, focusing upon values, consequences, etc. More importantly the approach enables holistic thinking, as ethical thinking is broadened, along with the development of the skills of reflection and learning, and the qualities and virtues of self-reliance and responsibility. The development of method enables the student to take responsibility for his own approach to ethical practice.

8.2.4. Values

While values can be seen as simply a part of any method, it is possible to focus on them, enabling the professional or student to develop a responsibility for them. A number of exercises can be used to facilitate this, which can be used in both professional and HE/FE contexts. These can be presented in the following order :

- Value clarification — this involves reflection on case studies or professional codes to make explicit the underlying values. It can also simply be achieved by encouraging individuals to reflect on their own values, along the lines of 'things which are most important to you now and in the future' (see Chapter 2).
- Value Classification — simply to hold values is of little use if it is not clear how they relate to the different parts of the person's life. Hence the group can be encouraged to look at the different values of the individual and classify them according to personal, professional or public (see Chapters 2 and 3, Fig. 2.3 and Table 2.1.). Such values still need to be tested to see why they are important and how they relate to each other.
- Value prioritisation — this exercise asks the group to rate the various values and make a priority listing. This may involve a list of groups of values (see Chapter 2, Fig. 2.4). The very process of prioritisation leads to attempts to justify the importance of the different values and principles.
- Value mapping — this begins to note how the different values relate to each other and to distinguish, for instance, general and fundamental principles from instrumental principles. For example, how far is the family an important value in itself, or an instrumental value which enables the development or maintenance of more fundamental values, such as of love, emotional stability, security, etc. (see Chapter 2, Fig. 2.5).

In reflecting upon professional values and principles, we quickly move to the principles of professional decision making outlined in Chapter 3,

Sections 3.3 and 3.4. Such exercises are an effective way of learning and also provide a clear way of examining the value conflicts which inevitably arise. They enable the learner to note shared values, the distinction between values and the possibility of effecting compromise, by accommodating several fundamental values.

The exercises also enable the individual to work on his or her own values and to test them at each stage. Such a process, precisely because it is so rigorous and learner centred, helps to develop several different qualities and virtues as well as skills, including

- the capacity to take responsibility
- the capacity to hold together very different and potentially conflicting values
- the capacity to respect values held by different groups
- the qualities of virtue and empathy, something developed by greater awareness of other's values
- intellectual skills of analysis, argument and justification
- greater personal awareness, as non-rational drives are distinguished from values and principles (see Section 2.3, Figs 2.1–2.3, 2.4–2.6 and Table 2.1).

8.2.5. Role-play
Role-play is a very effective means of focusing on values and can be divided into three phases

- study of and taking on the role, with particular reference to the aims and values of the character
- the play itself, where the values are tested in the dialogue and interaction
- the debriefing process, where the tutor and any observers can enable reflection on the issues raised and the further testing of values and ideas.

An effective role-play could be a bypass road enquiry, involving engineers, local community members, environmental and architectural activists, local authority leaders and so on. Such role-play could be built up from a local case and would require 'role cards' which give a brief history of the different characters and the basic values held by them. It would usefully involve a chairman who had some experience of such situations.

Role-plays can be of various lengths, ranging from two hours, which can be held in seminar rooms, to all-day events which are best carried out on sites away from the usual learning area. In any professional situation there may be a call for a clear reflective challenge which helps the different parties to understand what they are doing.

Approaches such as these thoroughly demystify ethics and strengthen the essential skills for ethical practice. It is possible for a class of reasonably intelligent undergraduates to develop a method and value map which is the equal of many of those contained in applied ethics textbooks.

These approaches also ensure that the public nature of ethical dialogue is learnt. The discovery, affirmation and development of ethical thinking is seen as arising from various levels of reflection and dialogue, not from a private world of thought or a distant repository of wisdom.

Both case study and role-play are made more effective with contributions from members of other disciplines and also representatives from industry. These can act as observers in the role-play and assist with the debriefing. They can also contribute to the reality testing of the various case study groups' responses. Such professionals can also model good dialogue and debate, and set up dialogue which can creatively challenge different perspectives. External figures can also be used as part of any assessment including the final assessment of group presentations.

8.3. Continuing professional development

8.3.1. Introduction
This group has significantly different needs, motives and expectations from undergraduate students. It also approaches the learning experience with different levels of experience, some of its members having already encountered the problems associated with an ethical decision in their work. They are therefore more demanding, complex and vocal students.

The programme will work to best advantage if the relevant experiences of the students can be incorporated into the programme. There is great benefit in encouraging the students to direct the 'form' of the programme within the context of a broad statement of intent, such as:

> This course raises and discusses moral and social problems that arise
> in the context of the design and construction of civil engineering
> projects. It will address the concepts embodied in the ICE Rules of
> Practice — including loyalty, integrity, values and principles — in
> making moral and professional decisions.

8.3.2. Course structure
Since all students on a course at this level should have experience of the problems caused by ethical dilemmas, it could be useful to begin by

encouraging participants to bring their own case studies to the course. These case studies would reflect problems that the students have encountered in their working lives.

Using these case studies the students could be asked to describe the decision making process they used in their cases and the outcome of those cases. Discussions and comparisons of these experiences could lead to a broad and differently viewed perspective of what it is to be 'ethical' and the background and consequences of a wide variety of cases.

The ICE Rules of Practice and other professional codes of ethics could then be introduced (see Chapter 3) and discussed, followed by an explanation of the ethical audit procedure set out in Chapter 6.

These considerations and discussions could be focused to encourage the students to derive a definition of *professionalism* and the *particular professional identity* of the engineer. As with undergraduate students, time should be spent at this stage exploring professional virtues and values. Role-plays designed to encourage reflection and use of these virtues should be used to demystify ethics and place the topic in the context of the engineering project.

8.4. Professional rules and codes

When consensus is reached, a further question could be posed. If all agree on the basis of a standard definition of professionalism and the values that underpin the professional, why then is it necessary to have a code of ethics? From this discussion could arise a definition of Professional Ethics and the need for written ethical standards (see Chapter 3).

As many engineers are now employed in professional teams of many different disciplines — architects, surveyors, economists, etc. — it will be useful to examine and critique other codes of ethics. The ICE Rules of Practice can be introduced and compared and contrasted with other professional engineering codes and those of other disciplines such as architects, managers, the caring professions, etc. If individual students or groups are asked to research and 'defend' these codes in open forum, the debate helps to develop greater understanding.

Having established the nature and content of a standard code of ethics, the remainder of the programme should focus on the investigation of specific civil engineering project based problems and case studies (see Chapter 5). The ethical audit methodology set out in Chapter 6, and the Falklands project referenced to the ICE Code in Appendix 1, can be used to highlight the various aspects of a working project. Students can then be asked to re-examine their own personal case study in the light of this methodology.

8.5. Extending the practice

The key element of ethical education is that of the development of the reflective practitioner. This means that ethical dimensions can be fed into virtually any aspect of practical training or professional development. The weekend training sessions which are often simply confined to the development of management skills and initiatives can include activities and cases which focus on the ethical dimension, and any work on skills or capacities can involve reflection on the ethical dimension. Where this kind of integrative learning is not possible, then greater emphasis could be placed on the use of the Internet, including the setting up of interactive case studies and of bulletin boards which would facilitate discussion and the sharing of good ethical practice. Such basic material could link into distance learning projects and to a professional qualification.[1]

---

[1] Robinson S. and Dixon R. The professional engineer: virtues and learning. *Science and Engineering Ethics*, 1997, 3, 339–348.

9

Future factors

> *One distinguishing feature of any profession is that membership in that profession entails an ethical obligation to temper one's selfish pursuit of economic success by adhering to standards of conduct that could not be enforced either by legal fiat or through the discipline of the market. Both the special privileges incident to membership in the profession, and the advantages these privileges give in the necessary task of earning a living are the means to a goal that transcends the accumulation of wealth: that of public service.*
> Supreme Court Judge Sandra Day O'Conner[1]

9.1. The professional culture — integration into the community

Ethical standards appear never to have been under greater attack than they are today. Whilst the ability to reason on a carefully considered basis, linked with professional integrity, independence and competence, may be the only way in which professionals and professional institutions can confirm their ability to contribute their special service to society, accusations of protectionism, of maintaining a closed shop, of complacency, are common. Professionals must be able to demonstrate that their contributions are based on specialist skills, not available to the lay person. This requires that professionals improve their presentation skills and demonstrate a transparency in the decision making processes.

---

[1] Http;//.nspe.org/eh1-lew.htm National Society for Professional Ethics.

Engineers are schooled in providing solutions to problems which are finite and based on technical logic. We should not be surprised to find that ethical dilemmas have no absolute answer and often cannot be logically quantified. Increasingly the need for a reflective capacity to be cultivated in the young engineer in parallel with analytical ability, is being recognised, and this is being addressed in the design of undergraduate and postgraduate courses. The reflective process is not unlike the skill that needs to be developed in creative design — an iterative, appraising procedure that enables many factors to be considered and related to one another without the use of mathematical formulations. The process is natural to all of us, but sometimes becomes buried in an education focused too precisely on quantifiable factors. Intuition need not always be a woolly, imprecise process, it can be incisive and accurate, if not confused by random emotions.

It is inevitable that we will meet ethical problems and conflict in our work. If these can be recognised as they arise and if we understand how to deal with them, accepting that there is probably no uniquely correct answer, then conflicts are avoided, and a balanced decision can be made, taking all factors into account. The future training and concerns of engineers must therefore develop this reflective capacity — which in itself makes our appreciation and enjoyment of our work and contribution much more fulfilling.

Action indicated
Considerable emphasis must be applied to the development and extension of coursework on ethical issues and the integration of engineering work into all aspects of community activity.

9.2. Global conflict

The growth in economic and political interdependence makes us all vulnerable to the instability of communities around the world. As nations adjust to technological and social developments, it is inevitable that these put great pressures on emerging and immature systems of governance, linked with the ambitions and inexperience of the leaders of these emerging and evolving societies. The more established nations themselves frequently seek to protect their own status by restrictive or manipulative practices, which generate conflict.

Stability is directly related to compromise, to adjusting to the behaviour of others, to the willingness to make personal sacrifices in serving some greater need — whether of family, nation, mankind or creation. With the ability to develop an objective, but compassionate, understanding of the value of toleration, it is possible to avoid the extremes of anarchy and totalitarianism. The role of the United Nations (UN) in the monitoring and development of basic human rights, and their relationships to human duties, is of great importance when the national government may not yet have evolved a system which provides such direction and guidance.

The engineer's ability to serve the world provides an opportunity to contribute to the stability of the emerging nations, but also exposes him to the conflicts and uncertainties of the unstable state. The ability to handle this kind of conflict is becoming an essential component of the engineer's competence, and there is a need to ensure that the history, culture and aspirations of those people among whom engineers find themselves working is well understood. There is also a need to seek out and benefit from the experience and understanding of other specialists — theologians, historians, economists, lawyers, politicians, sociologists, geographers, etc. Such a preparation is not only invaluable, it should be taken by the engineer as being a duty, enabling him to understand problems which may not arise in his own culture.

Iranian appointments

The telephone is unreliable, I cannot make an appointment to see the Head of the Planning Department.

Why must you make an appointment? Go and see him. Our custom of hospitality requires that he see you if you turn up, even if you have to wait for some time.

Ethical meaning and understanding are developed out of dialogue, both within and outside the profession. While there are 'big principles' shared by many different groups, these principles are not always known, or articulated. Hence, any ethical concurrence of meaning requires dialogue. We must not only know how to communicate, but must be prepared so to do. This may mean making a choice between several different outcomes none of which is 'wholly good'. It may also require relating very different values which are good or right in themselves, but which appear to be in conflict. Shared reflection enables the different ways of handling such ambiguities to be worked out in situ.

The savouring of governance by the salt of professionalism can make a national dish more wholesome and appetising. The qualities of integrity and concern for the public benefit contained in the first two of the ICE Rules of Practice are essential in responding to needs in an unstable environment. It also helps to like people.

135

Action indicated
It is important to pursue the seeking and sharing of experience and understanding in international affairs and social behaviour. The establishment of a forum in which the exchange of views could take place would be of great help — perhaps the professional organisations could collaborate on such a venture. An awareness of the activities of the UN organisations and agencies would also be of help.

9.3. The nature and culture of work

Changes are taking place in the nature and culture of work and employment. There are significant changes in conditions of engagement — short-term contracts are becoming more widespread, partnerships are giving way to the incorporated companies, flexible working hours are not uncommon. These changes, whilst giving more freedom to both individual and company, make it more difficult to develop corporate loyalty and a corporate culture which includes ethical components. It also discourages the development of mentoring relationships, including the availability of retired members of an organisation to contribute from their experience to the development and understanding of succeeding generations.

Whilst teamwork is being emphasised and practised, it may not be easy to develop teams whose members have mutual respect for one another and where the young practitioner can be instructed and tested. The ethical response is rarely one that is or should be, individual. The more that data is shared the more it is useful and the more that responsibility is shared the more an ethical response becomes possible. Ethics is not simply about framing judgements but seeing how the good can be made possible in practice.

An ethical response cannot be worked out without clearly determining responsibility. The changing nature of work, with short-term or contractual relationships between individuals rather than long-term commitments, can reduce the sense of responsibility — 'after all I shall not be here when this is built or commissioned'. It must be realised that responsibility is a personal acceptance of a role, not an obligation to some organisation or employer.

The Rules of Practice of the ICE acknowledge this individual responsibility as being part of the engineer's own duties. On certain projects, the particular responsibilities of the team members must be established and accepted at the outset of the work.

Action indicated
The professional institutions must recognise these changes in loyalties, and ensure that their members understand their own personal responsibilities. They should establish points of reference within their organisations to enable practitioners to consult other experienced engineers when faced with ethical dilemmas.

9.4. Engineering technology
Rapid changes in technology may generate new ethical problems which will need to be addressed by both individuals and by the professional organisations. The growth of IT, with instant access to much more information, provides both opportunities and problems for the busy practitioner. How much of the information available should be consulted? Is it accurate? Is there a need for some form of professional refereeing of available data to provide a considered and reliable database? Is this a task for the professional organisation — perhaps on subscription?

The success of engineering projects and developments has, paradoxically, generated many new problems, and these will no doubt become more urgent in the future — transport, energy consumption, energy generation, fossil fuels, non-recoverable consumption of materials. Much attention is being paid to the problems of the recycling of waste products, of designed-in recovery procedures, as well as to the problems of pollution control.

These are all the subjects of continuing and growing discussion and debate. It is reasonable to assume that the same ability that developed the facilities which are now seen to pose problems, could be applied to the resolution of those problems — and there is evidence that this is taking place. This is an ethical issue as well as requiring high technical ability. Care must be taken to ensure that the solutions avoid becoming problems themselves at some time in the future.

Action indicated
Private and public research and development initiatives aimed at ensuring a full understanding of the implications of new technologies must be encouraged and promoted, with effective communication of results of such work and the preparation of guidelines for the engineer.

9.5. Conclusions
The bottom line for professional ethics is the human good. How do our actions affect those involved, either now or in the future? Ethics is not

a bolt-on activity, to be adopted or not, as we wish. It is part of the identity of the professional engineer, central to thinking and to producing. Professional ethics is value-centred not value-added.

There is, and must continue to be, a growth of internationally acceptable global values. It is useful to note that all cultures seem to be based on the premise of 'do as you would be done by', which indicates that ethical solutions can transcend cultural barriers. There is a universal expectation of the possibility of human happiness, of fulfilment, of an equality of opportunity. To contribute as we can to the well-being of mankind and the husbanding of natural resources requires trust, compassion and sincerity — all natural characteristics of the human being. Perhaps our only duty is to be ourselves.

In the words of Sydney Evans of Salisbury, late Dean of Salisbury

To be is to have duties

Appendix 1. Case study: Mount Pleasant Airport, Falkland Islands

A1. Introduction

The primary aim in constructing the airfield at Mount Pleasant was to safeguard the security of the Islands by providing a facility to enable the rapid reinforcement of the garrison in the event of a crisis. The resident garrison could then be reduced in size.

The consequences of such a project are, however, far-reaching and needed to be fully considered in its planning and execution. Here, an audit of the several stages of this particular project are considered — as described in general terms in Chapter 6 — setting out the most significant aspects of the project with respect to the ethical consequences of the decisions taken at each stage. The integrating and conflicting factors likely to be present at each stage are itemised in the various tables in this Appendix, and are of particular importance.

A1.1. The nature of the project

A1.1.1. The location

The airfield was sited some 60 km from Port Stanley, requiring the construction of a new road to link it to the port facilities. This isolation helped to protect the community of Port Stanley from the major impact of the very large workforce. A new port facility was developed very close to the airport by permanently mooring a ship nearby which had been used to take out the first of the materials and construction workers. This ship was later replaced by a permanent berthing facility. Fig. A1.1 shows the planning factors considered in designing the facilities associated with

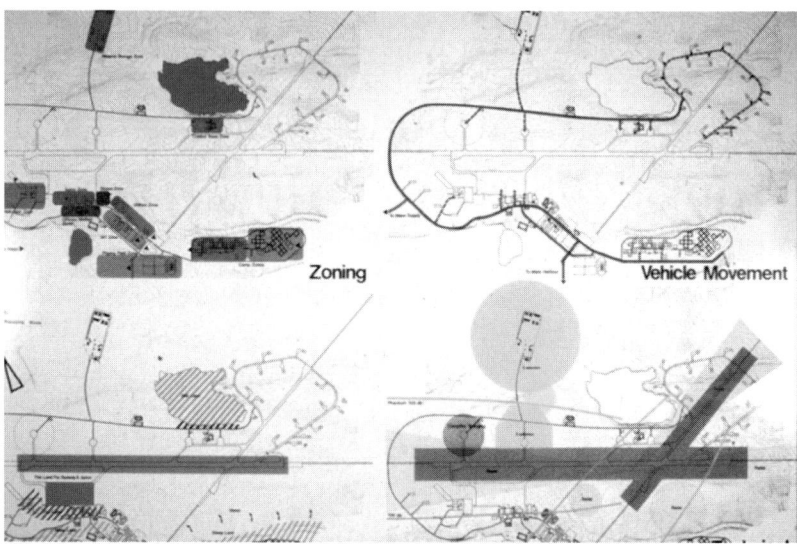

Fig. A1.1. Planning factors considered in the design of the facilities associated with the airport

the airport. These facilities included temporary and permanent residential accommodation much more extensive than those at Port Stanley itself.

The impact of the construction work was considerable. For example, the quarries were the most significant single development that had ever taken place in the Islands. Steps were taken to ameliorate the worst effects of the development upon the environment, and to create a self-contained way of life within the construction and operational communities. Perhaps the contrast between the natural habitat and the aspirations of human endeavour are illustrated by the reflections on the natural harbour at Port Stanley, and the celebration of the achievement of the construction team in the award made to the contractors by a leading construction journal in the UK.

A1.1.2. Habitat

Most of the land is undeveloped moorland peat bogs, with outcrops of rock and scree. There are many small and shallow lakes, but few major streams. There is a very long and indented coastline. The native flora and fauna are limited in variety, but include some rare and interesting species of South Atlantic life forms. Penguins and other seabirds are very common, and there are colonies of breeding elephant seals in the surrounding areas.

Plate 5. Penguin colony on the Falkland Islands

A1.1.3. Social aspects

The Airport project had a profound effect upon the social life of the Falklands community. The population of the Islands was just under 2000 people at the time of the conflict with Argentina, of whom about 900 lived in the largest town of Port Stanley. Their chosen and preferred way of life centred around sheep-farming, with a few additional service jobs and a little fishing. The way of life was similar to that of some of the islands off the coast of Scotland. Leisure activities were simple and communal, involving the whole family, as illustrated by the regular Saturday evening dances and social events which were attended by all generations.

The natural resources of the Island are limited, little cultivation having taken place. The community was dependent upon imports for most of the commodities needed for its daily life. The general level of incomes was low, but adequate for the needs of the Islanders. It is of interest to note that a consultancy practice of, say, 1000 staff together with their dependants outnumbered the whole population of the Falklands Islands.

It was to this sheltered community that more than 3000 workers came to build the airport, affecting all aspects of the community's way of life.

Plate 6. LMA's process area (now ADR) on the Falkland Islands, 1985

A1.1.4. Range of influence

How does this project compare with the scale of values discussed in Chapter 2? Clearly it has had an adverse effect on the individual Falklander's desire for a quiet and simple way of life. Removing the airport from the immediate vicinity of any settlement has had the effect of reducing this impact, but some permanent adjustment will be necessary. The presence of the airport will make the Islands more attractive to others, so some greater individual satisfactions will also arise. The family life of the residents is influenced by the new facility. A bigger range of employment is now available, with a wider variety of social amenities. The sports and recreational accommodation provided at the airport will be available to all the Islanders, and better transport within the Islands is possible.

The wealth of the whole community will be increased, but will need to be carefully monitored to ensure that the economy remains in balance. The community is in much closer touch with the rest of the world than had previously been the case.

It is not unlikely that in the future the development of some of the natural resources of the South Atlantic will become financially viable. In this event the availability of the improved transport facility will be

Plate 7. Pleasant Peak Quarry B, Falkland Islands, 1985

very valuable, not only to the Falklanders, but to the world community at large. It will, however, also encourage, perhaps adversely, the exploitation of the resources, and care will need to be exercised to ensure that this is not overdone. In considering the ethical issues involved in this project it will be helpful to work through the stages of the development and evaluate them against the ICE Rules.

A2. The decision makers

We start by recognising the value systems of the several parties and their personal and professional interests. These include institutional and legal codes and laws. Those factors which are likely to impact on the project, including all the factors arising during the lifetime of the project, need to be discussed by all parties so that some convergence into an agreed overall brief can be prepared. At this stage we need to note the differing 'missions' of the several parties — defence, treasury, security, confidentiality, commerce, local community, contractors, suppliers, consultants, workers, ecologists.

These might include speed of construction, cost control, improvement of trade, disturbance to the peaceful environment, effect on farming,

Plate 8. Completed airport at Mount Pleasant in the Falkland Islands

changes in social structure, profitability, logistics, reduction of risk (commercial), enhancement of reputation, work prospects, higher rates of pay, living conditions, and the effects on indigenous flora and fauna.

Table A2.1 sets out the key decision makers and their interests, with examples of the types of ethical issues that can arise, together with the integrating and conflicting factors contributing to, or inhibiting, progress towards the conclusion of a satisfactory project. Reference is made to the Rules of Practice of the ICE set out in Table 3.4, and to the general comments on these rules set down in Chapter 3.

A2.1. The client

As with all major public sector projects, several ministries were concerned with the project. In this case this included the Foreign Office, the Ministry of Defence, the Department of the Environment by way of the Property Services Agency (PSA), and, of course, the Treasury. Professional engineers and others were involved with, and employed by, the government departments and the private sector consultants and contractors.

The aims of the client departments were to achieve an acceptable operational airport, for both public and defence use, to demonstrate an openness in the international implications of the project, to reassure the Islanders that their affairs were being considered and that the quality

Table A2.1 (pages 145–147). The Falklands project ethical audit: the decision makers

Key decision makers	ICE Rule references	Ethical considerations
The Client		
Ministry of Defence	**1, 2, 4, 11**	The safety of the Islands and
Foreign Office		their population
Property Services Agency		The appropriate control of public
Political acceptability		expenditure
Economy		Diplomatic relationships with
Image		neighbouring states
Value for money		The public perception of value
Interests of workforce		The well-being of residents and
Use of available resources		workers
		The proper use of natural resources
Consultants		
Reputation	**1, 3**	The well-being of their staff
Interests of staff	**4, 5, 11**	Their international reputation for
Professionalism	**2, 7, 16**	the adequate design and control of
		such an unusual project
		Openness and good relationships
		with all parties
Health and safety	**3, 4**	Clarity of definition of all
		responsibilities
		Care for the resident community
		and for the environment
Contractors' and suppliers' interests		
Minimal alterations	**1, 3, 4, 11**	Clarity of contractual obligations,
Enhanced reputation	**2, 7**	the avoidance of doubt
Interests of work force		Full understanding of constraints
Minimal costs/maximum profits		
Compensation for extra work		
Labour and material resources		
Health and safety in transit		
and on site		
Enhanced reputation		
Interests of work force		

Table A2.1 continued

Key decision makers	ICE Rule references	Ethical considerations
Special interest groups		
Employment opportunities	**2, 3, 7, 11, 14**	Need to maintain Island's workforce for Island's needs
Environmental impact		Need to note and protect rare and important wildlife habit
Falklands Government		
Pollution		Need to maintain good relationships with Island Government
Sport/recreation		
Amenity		Collaboration with local farming/fishing interests
Local industry — farming/fishing		Sharing of recreational facilities
Security, policing		Collaboration with local police
Employees		
Career progression	**1, 3,**	Good working conditions
Job satisfaction	**4, 5, 11, 13**	Health and safety, availability of resident medical attention
Working conditions	**2, 6, 7, 16**	Adequate compensation
Pay		Recreational facilities
Integrating factors		
Communication/negotiating skills		Track record of success in similar projects
Transparency in transactions	**4, 11**	High standards of integrity
Common goals/trade-offs	**2, 7, 14, 15, 16**	Ability to work as a team
Team-working		Mutual respect and the recognition that the contribution of all parties was essential — selfishness and covert exploitation of the difficult conditions would have been very detrimental.
Definition of responsibilities		
Acknowledgement of particular contributions		
Acknowledgement of authority —		Senior representatives of all parties — both in the UK and in the Falklands — had proven and well-defined authority and integrity
extrinsic/intrinsic		
Risk control		

Table A2.1 continued

Key decision makers	ICE Rule references	Ethical considerations
Conflicting factors		
Divergent interests	**4, 11**	Pressures of programme and
Secrecy		the logistical problems of
Incompetence		transporting all staff and
Lack of recognition of		materials from the UK created
authority		occasional conflicts of
Conflicting interests —		interest. Security in camp and
commercial, personal		in transit, with the need for
Risk exposure		political discretion, also
to manage the high risks of		required careful management.
exposure in the pioneering		Special care was needed
environment		

of their lives would not be reduced by the provision of these extended transport facilities, and to obtain value for money and full accountability. It was a condition of the contract that only UK consultants and contractors would be engaged, and that materials and supplies would all come from the UK. Some aspects of the project were subject to security restrictions, requiring the exercising of discretion by those involved.

These aims are variations on the general aims given in Table 6.1, taking into account the particular nature of this project. They would all form part of a project ethical manual.

A2.2. The consultants

It was the task of the consultants and contractors involved to satisfy all those requirements described in Section A2.1. The reliability of the advice of the consultants to the client — the contractual relationship was with the PSA — depended very much upon the integrity and independence embodied in Rules 1, 3, 4 and 11.

Rule 1. A member shall discharge his professional responsibilities with integrity and shall not undertake work in areas in which the member is not competent to practise.

Rule 3. A member shall have full regard for the public interest, particularly in relation to the environment and to matters of health and safety.

Comment
The adoption of UK standards for contractual procedures and standards ensured that the requirements of clauses 4, 13 would be met.

Rule 4. A member must understand and comply with the Laws of the communities within which he practises and with International Law. Where National professional Codes exist, these should also be followed. Where neither Laws nor Codes exist, then the Institution's Rules of Professional Conduct should be followed.

Rule 11. A member shall, consistent with safety and other aspects of the public interest, endeavour to deliver to the employer or client cost effective solutions. A member shall not comply with any instruction requiring dishonest action or the disregard of established norms of safety in design and construction.

Comment
Because speed was of the essence in this project the client sought out practitioners with proven prior experience and competence in the type of construction works involved. This involves consideration of Rule 7.

Rule 7. A member, either individually or through the member's organisation as an employer, shall afford such assistance as may be necessary to further the continuing development of the individual and of other members and prospective members of the profession in accordance with the recommendations made by the Council from time to time.

Comment
Clearly on such a high speed project the need for the professionals to have the confidence of the client and to respect each other's integrity and expertise was paramount. It was essentially a major team project, bringing together as partners organisations that had not previously collaborated, but all of whom knew and trusted the quality of the work of the others.

A2.3. *The contractors*

All the contractors involved in the project were very experienced in operating in remote locations, and were known to have the resources needed to carry out the work. They were also experienced in public sector work, and the procedures needed to reduce to a minimum misunderstandings and potential conflicts. The need for public account-ability was also appreciated.

Pre-tender briefings and site visits helped to ensure that there was a full understanding of all restrictions on material and labour resources, and of the health and safety requirements for this unusual project. This careful preparation helped considerably in securing full collaboration between all the parties as members of a very interdependent work-force.

A2.4. *Special interest groups*

Working procedures between the project team, particular the on-site management, were set up to ensure that a full understanding between the construction team and the local interests was established and maintained. These included the farming and fishing interests as well as the local security services — established garrisons and the police force. The impacts upon local amenities, recreation facilities, employment and resources, were also carefully considered. Regular meetings between the project management team and the local representatives took place.

A2.5. *The employees*

The needs of employees of the construction team were also carefully considered. This applied to all employees — the PSA, consultants, contractors, subcontractors and suppliers. On such remote and isolated projects the responsibilities of the management team for the well-being and happiness of the workforce become of particular importance. Opera-tives rarely could leave the project to lead a life of their own, as would be the case on a home-based project. Tour periods were of about six months' duration, almost entirely on a bachelor basis.

The importance of job satisfaction, the maintenance of career develop-ments for the professionals, and the balance of the rewards by the differing employers, was well recognised. The provision and operation of common residential and recreational facilities, and health and safety facilities, were also important factors in ensuring the smooth running of the project.

A2.6. *Summary notes*

Traditional construction projects involve adversarial behaviour and conflict which arise out of the differing objects of the various groups involved. The role and responsibilities of the engineers have already been explored in Chapter 3, but it must be recognised that an engineer may also operate in the capacity of client, consultant, contractor or supplier of goods. Each of these roles carries its own separate attitudes, responsibilities and aspirations. In the context of the project team of client, consultant and contractor, various operational and ethical conflicts may arise.

In the case of the many people involved in the Falklands project, conflicting interests were likely to be exaggerated by the very unusual, and very challenging, conditions as described in Section A1. It was therefore essential that all parties worked together in an atmosphere of mutual trust and openness. This was recognised at the outset of the project, and great care was taken to select an experienced team, whilst acknowledging the need to ensure the full accountability of all parties for the maximum economy to be achieved, for the target dates to be met, and the specified quality of facility delivered. The establishment of good communication facilities between the Islands and the UK was essential, and frequent meetings of all members of the senior management team were held, so that all problems were dealt with as they arose.

Commitment to the single aim of realising the client's requirements for a sound operational airfield within a short timescale was very evident, and a management structure developed which related all partial or short-term objectives to this aim. Care was taken to ensure that all requirements were fully spelt out, all decisions recorded and verified, and every effort made to reduce the stress level imposed upon all staff and operatives.

A clear understanding between the several decision makers as to their several levels and areas of responsibility and authority was developed.

A3. The brief

The key areas in which ethical issues might arise during the brief-making stage are set out in Table A3.1. The particularly unusual features of the project were the remoteness of the site — with the need to simplify the complicated logistical problems, to allow for the staffing problems of travel, isolation, local care, policing, etc. — and the need to make a minimal impact upon the rather fragile local ecosystem, both human and ecological.

For political reasons it was required that all personnel be UK nationals, and all materials be of UK origin. Only concrete aggregates were obtained locally, and these required the development of new borrow pits and quarries, with significant environmental constraints.

Table A3.1 (pages 151–152). The Falklands project ethical audit: the brief

Factors	ICE Rule references	Ethical considerations
Economic factors		
Public accountability	**1, 3, 4, 5**	Risk assessment, communication of hazards (e.g. land mines)
Value for money		
Low running costs — energy balance		Energy audit
Cash flow/time constraints		Equitable contractual procedures
Early start on site		
Space		
Adequate space standards	**4, 11**	Security of quality, durability
Arrangements of space, facilities required for long-haul flights	**2, 7, 14, 15, 16**	Quality of life for users in adverse conditions
Design standards		Economy of space, reduction of impact
Preferred materials, services local/imported		Amenity, social, cultural needs, e.g. chapels, clubs, recreation
Communications		Health care, emergency procedures
Maintenance management/costs		Non-toxic, low-energy, recyclable materials
Public space		Security of communication systems
The site		
Alternative locations	**5, 7, 11**	Environmental impact assessment
Conservations		Alternative sites investigated (see Section 6.4)
		Land use

Table A3.1 continued

Factors	ICE Rule references	Ethical considerations
Legal factors		
Planning conditions	**4, 11**	Lack of external authority requires self-regulation and responsibility
Responsibilities		
Liabilities		Lack of experience of local community
		Equitable contractual conditions
		'Open-book' accounting procedures
		Managerial integrity, close liaison
Contractual procedures		
Selection procedures for contractors	**4, 14**	
Competitive bids		
Manage/design/build		
Private finance		
Insurances		
Programme		
Integrating factors		
Clarity of definition of brief	**7, 11**	Co-operation of all parties
Itemisation of elements		Integrity, independence, impartiality, competence, discretion
Mutual understanding		
Team experience		
Conflicting factors		
Confusion of meaning	**11, 14, 15, 16**	Long-range communications leading to possible misunderstandings
Assumptions of meaning		Intensity of design/ procurement programme generates stress
Confusion of wants and needs		Lack of adequate information

The assembled teams of client representatives, consultants, contractors and suppliers were all very experienced, and in many cases had collaborated previously on other public sector projects. An input from all the decision makers was required in the preparation of the brief. Many of the construction elements had to be fabricated in the UK and shipped out, to help to limit the size of the resident labour force.

In spite of this close collaboration it was essential, in such a high-profile project, that full public accountability be maintained to ensure the good husbandry of public resources and demonstrable value for money.

The timescale for the project was also short. For example, a period of only about ten weeks was available for the production of contract documents for approximately $200 million of work

The brief included a full environmental impact analysis by an independent consultant, which set standards for the control of the site and the care of the surrounding open areas. High security standards on information flow, personnel, and communication procedures were specified.

In spite of these difficulties or perhaps because they created a heightened awareness of the problems, the brief and design were carried out with a minimum of misunderstanding. The professionalism of all operatives was of a high standard.

A4. The context

Table A4.1 sets out the particular characteristics of the context of the Falklands project. In this case, the political, economic and social factors required greater attention than the physical environment. While the physical factors were of great importance, they were relatively clear and simple to define. The optimal solutions arose naturally from the nature of the context, and they were agreed and understood by all parties from the outset of the project and incorporated in the brief and contract conditions.

Because of the unusual nature of the project, attention was focused on the major issues, and they were fully discussed before the start of the contract. This made the definition of procedures and the management of the site operations a recognised part of the regular progress reviews throughout the project. Paradoxically, the more difficult the situation the more attention it attracts, and hence the greater possibility of positive control. Those projects which take place in more 'normal' contexts may suffer from the neglect of significant, but commonplace factors, which can be assumed by all the senior management team to be resolved without

Table A4.1. The Falklands project ethical audit: the context

Factors	ICE Rule references	Ethical considerations
Social context		
Employment — limitations on use of local labour	**1, 3, 4, 11**	Impact very considerable
Social services — none available		Need to minimise contact
		Disciplined control of expenditure
Economy		Self-regulation
Housing — none		Contractual cash flow exceeds
Safety		Islands' total budget
Pollution		Site policing very important
Physical context		
Access/transport	**3, 4, 7, 11**	Limitations on communica-
Materials, availability		tions with community
Services required (water, power)		Understanding of impact considerations — use of local
Amenity		materials affects local ecology
Conservation		Minimum profile for all
Sustainability		operations
Energy consumption		Almost all materials and
Topology		energy have to be imported. There are no local suppliers
Integrating factors		
Mutual benefits — Island/UK	**1, 3, 4, 11**	Good communication system
Improved communications	**2, 7**	with local representatives
Thorough and public investigations		Social gains of improved facilities — transport,
Wide consultation		infrastructure, amenities
Involvement of community		
Conflicting factors		
Divergent interests	**8, 14, 15, 16**	Lack of understanding
Inadequate information		Carelessness, thoughtlessness
Covert investigations		Uninformed decision making
Change of land use		Arrogance
		Poor liaison with local administrators, short timescale for decision making

discussion. Communication and agreed aims are again the key to harmonious relationships.

A5. The design

In considering the implications of the design-stage decisions in the case study of the Falklands project, many unusual factors were taken into account. The total context of the project, because of the very small economic 'ecosystem' of the Islands, had a considerable effect upon the community. The commercial and management aspects of the project were very important. The key decision was to insulate the local economy from the project insofar as this was possible.

The project was located some 60 km from the major settlement at Port Stanley, and the access road between the construction camp and the community was not upgraded until late in the construction period. Since for both political and practical reasons all materials and labour were brought from the UK, a separate handling facility was needed. The excellent harbour at Port Stanley was not available, being inaccessible from the construction site, and so a temporary harbour was created by mooring one of the first supply vessels, with adequate handling equipment, at Mare Harbour, about 7 km from the selected airfield site. This acted as a jetty and all subsequent cargoes were unloaded across its deck, which was connected by a bridge to the shore. There was thus little impact upon the port facilities at Port Stanley, which had already been considerably affected by the garrison stationed there following the conflict. One of the major reason for the construction of the airport was to enable the garrison to be reduced considerably in size.

For similar reasons there was a need to ensure that subsequent maintenance of the facilities was reduced to a minimum and capable of being undertaken by a local team of operatives, without access to sophisticated equipment or the ability to purchase any necessary materials from the local community. Prefabricated timber buildings were adopted for the residential buildings, and simple steel framed buildings for the workshops, power house, aircraft hangers, etc. When the logistics, programme, or financial constraints on projects are particularly difficult, then the use of tried and tested construction techniques and materials is usually preferable.

Since the future development or use of the airport was uncertain, allowance had to be made for possible major changes of use, and the removal of facilities might be necessary.

The Falkland Islands are renowned for the unusual variety of wild life and as breeding grounds for some unique and significant birds and marine mammals. An independent, thorough, environmental impact analysis of the project was commissioned and carried out in parallel with the design process. This enabled certain important breeding grounds to be identified. It was also a requirement that areas be identified for the extraction of aggregates and the opening of quarries to provide materials for concrete and for the surfacing of pavements.

It was also necessary to locate an adequate water supply for the project using local streams, recognising that the operational demand would be less than the construction demand, the provision had therefore to be easily reversible on completion of the project.

All these special factors were considered in design team sessions (see Table A5.1). Projects of any size have their own special characteristics, to which due weight must be given during the design process. The professional engineer may, indeed, not be aware of the ethical factors which permeate his decision making, forming as they do part of the culture of decision making for civil engineers. An ethical audit makes these factors more visible, and enables them to be given due weight in the process, avoiding conflicts that might arise otherwise at a stage when it is difficult or impossible for them to be resolved. Again the avoidance of conflict depends upon good communications, upon the agreement of common aims, and upon the trust between the several parties involved in the project.

The completed designs were officially 'signed-off' by the design team and client, and described in a definitive specification and well-presented schedules and drawings. A reference copy was carefully filed, so that any subsequent changes that might prove unavoidable could be adequately cross-referenced to the original proposals.

A6. Implementation

The general descriptions given of the project have drawn attention to the special conditions applicable to it. Because of the special circumstances involved, the need for close liaison between members of the project team and the local community was recognised. Considerable authority was delegated to the site staff — both supervisory and contractual — for external and internal relationships. The close involvement of the site staff in the earlier stages of decision making, and the close and continuous communication between site and home-based teams — clients, suppliers, consultants and others — ensured that decisions could be made which did not conflict with the strategic criteria of the client body in the UK.

Senior level communication was established with the local government and community officers to ensure that external relationships were good. Where possible, the sharing of facilities was arranged, thereby enhancing

Table A5.1. The Falklands project ethical audit: project design

Factors	ICE Rule references	Ethical considerations
Design proposals		
Alternatives		Avoidance of impact on local
Optimisation	3, 4, 5, 11,	economy
Maintenance	2, 7	Sound, easily 'buildable'
Engineering adequacy		design, free form, high-tech
Materials' choice		elements
Relevance to context		Robust, adaptable designs
Aesthetics		Low profile, environmentally
		compatible designs
Impact		
Long-term effects	2, 3, 4, 7, 11	Flexibility for alterations
Reversibility		Need for full impact
Sustainability		assessment
Environmental impact		Local wildlife conservation
assessment		Design for easy removal/
Short-term impact		disposal of temporary works
disturbance (light/sound)		
pollution (effluent/dust,		
etc.)		
Integrating factors		
Well-documented proposals	7, 11	Low impact project,
Objective assessment of		well-documented and
alternatives		discussed
Objective and balanced		Agreement with Island
impact assessment		Government
		Sensitivity of local staff
Conflicting factors		
Partial or confusing	5	Confrontational staff
presentation		Lack of care in appearance of
Prejudiced assessment of		buildings and ancillary works
alternatives		
Insensitive designs		
Extravagant use of materials		
and energy		
Short-term objectives		

Table A6.1. The Falklands project ethical audit: implementation

Factors	ICE Rule references	Ethical considerations
Team members		
Client's role	**1, 3, 4, 11**	Relative responsibility of site management and HQ management
Consultants	2	
Contractors		Relationships between client, consultant, contractor and local community
Suppliers		
Falkland community liaison		
Control procedures		
Quality assurance	**5, 7**	Establishment of independent quality and testing facilities with clear responsibilities
Testing procedures		
Progress control		Agreed progress/cost reporting procedures
Construction management		
Quality control	**3, 5, 11, 13**	Major role for logistics officer
Materials control and supply		Autonomy of site staff
Storage on site		Disposal of materials and plant on completion
Plant and equipment		
Access to site		
Temporary harbour		
Labour management		
Site hygiene	**3, 4, 5, 11**	Welfare and transport control
Health and safety	**2, 7**	Security/policing arrangements
Labour relationships		Local medical care facilities
Adequate training in site procedures		
Provision/maintenance of personnel accommodation		

Table A6.1 continued

Factors	ICE Rule references	Ethical considerations
Integrating factors		
Regular all-party meetings	**4, 11**	Site liaison meetings and
Accurate reporting on cost	**2, 7, 16**	agenda
and time control		HQ (London) liaison with
Accurate and objective		site management
reporting of any contingencies		Regular, formal reporting
Communication at all levels		Transport liaison
Close liaison of site		
management and quality		
control teams		
Conflicting factors		
Adversarial attitudes	**3, 11, 13**	Isolation of site personnel
Withholding of information	**7, 16**	Indifference of absentee
Poor site cleanliness and		project management
safety controls		Confusion as to
Lack of authority at		responsibilities
construction site level		Overstressed site management
Interference from client/		Indifference to local
consultant in construction		population
management		Limited site management
Poor relationships with local		priorities
community		

the opportunities for the improvement of community activities. Particular attention had to be given to the provision of accommodation and recreational facilities for the site staff, and morale was maintained at a high-level throughout the project. Almost all the Rules of Practice are relevant to this stage of the project (see Table A6.1).

Construction workers frequently take a great pride in their work, and it was observed on the arrival of the first aircraft on the airstrip that many of the operatives lining the runway to welcome it were weeping with delight at their achievement.